Advances in Spatial Science

The Regional Science Series

This series contains scientific studies focusing on spatial phenomena, utilising theoretical frameworks, analytical methods, and empirical procedures specifically designed for spatial analysis. Advances in Spatial Science brings together innovative spatial research utilising concepts, perspectives, and methods relevant to both basic science and policy making. The aim is to present advances in spatial science to an informed readership in universities, research organisations, and policy-making institutions throughout the world. The type of material considered for publication in the series includes: Monographs of theoretical and applied research in spatial science; state-of-the-art volumes in areas of basic research; reports of innovative theories and methods in spatial science; tightly edited reports from specially organised research seminars.

The series and the volumes published in it are indexed by Scopus.

More information about this series at http://www.springer.com/series/3302

Kieran Donaghy • Arash Beheshtian •
Ziye Zhang • Benjamin Brown-Steiner

The Co-evolution of Commodity Flows, Economic Geography, and Emissions

Springer

Kieran Donaghy
City and Regional Planning
Cornell University
Ithaca, NY, USA

Arash Beheshtian
Altum Group Advisors
New York, NY, USA

Ziye Zhang
Paul and Marcia Wythes Center
Princeton University
Princeton, NY, USA

Benjamin Brown-Steiner
Arlington, MA, USA

ISSN 1430-9602 ISSN 2197-9375 (electronic)
Advances in Spatial Science
ISBN 978-3-030-78554-3 ISBN 978-3-030-78555-0 (eBook)
https://doi.org/10.1007/978-3-030-78555-0

This Springer imprint is published by the registered company Springer Nature Switzerland AG.
The registered company address is: Gewerbestrasse 11, 6330 Cham, Switzerland

Preface

An oft-quoted observation attributed to Lao-tse and/or Mao Zedong is that a journey of a thousand miles begins with a single step. The single step with which the long intellectual journey of this book was begun was a presentation at the National Urban Freight Conference in Long Beach, California in February 2006 (Donaghy 2006a). There I presented a paper setting out a New Economic Geography (NEG) model of a spatial economy embedded in a commodity-flow model of recent provenance, suggesting that at some time in the not too-distant-future such a model might be operationalized to examine the co-evolution of commodity flows (freight movement), economic geography, and atmospheric emissions. This work was motivated in part by an invitation to participate in a contemporaneous project of the University of Illinois's Center for International Programs and Studies to examine how globalization—the increased movement of goods, capital, information, and people between locations within and between countries—was affecting people and the environment (Donaghy 2006b). The importance of this subject was later reinforced in 2009 by a workshop at the University of Paris on the history of environmental imprints of cities and led to the drafting of the material published in Donaghy (2012) and presented in Chap. 1.

Further developments of this modeling framework presented at a meeting of the International Society of Dynamic Games in Sophia Antipolis, France in July 2006, an INFORMS meeting in Pittsburgh, Pennsylvania in November 2006, and a workshop on complexity in Barchem, the Netherlands in September 2007 advanced the theoretical specification (Donaghy and Sheffran 2006; Sheffren et al. 2006; Donaghy 2007, 2009). In 2009, the U.S. Environmental Protection Agency (EPA) "Science to Achieve Results" (STAR) program solicited research proposals to examine, inter alia, how atmospheric emissions and air quality would evolve if the trend of increasingly transport-intensive globalization continued into the future. A grant from the U.S. EPA STAR program provided the resources to begin the applied empirical work in earnest. The combined efforts of Gianfranco Piras and Jialie Chen, under the guidance of Randall Jackson, contributed to the development of a regional econometric input–output model (REIM) whose estimation would give rise to commodity-flow data (see Donaghy and Chen 2011, Chap. 3) with which structural-equation modeling could later be conducted. The 31-annual-observation time-series on commodity flows between 13 industries in 13 states of the Midwestern and Northeastern regions of United States (U.S.) and the rest of the country is the first such time-series of their kind of which we are aware.

Benjamin Brown-Steiner made good use of these data in his dissertation research to examine the co-evolution of freight movement, economic geography, and associated non-point-source black carbon emissions (see Brown-Steiner et al. 2015, 2016; Chaps. 4 and 5). These data are available from a Springer website associated with this book.

While the data derived from the REIM modeling enabled the co-evolution of the commodity flows, economic geography, and atmospheric emissions to be *tracked*, the data themselves offered no behavioral explanation of these phenomena; hence, there was a need to confront a theoretical structural-equation model with our empirical data, estimate its parameters, and evaluate its fit. The perception of this need led to further explorations in the specification of dynamic commodity-flow models and their possible use in dynamic-game simulation experiments, ably supported by Arash Beheshtian (see Donaghy 2017; Chap. 6).

The econometric estimation of a nonlinear continuous-time mixed-sample 13-sector model of commodity flows between states in the Midwestern and Northeastern regions of the U.S. and the rest of the country was begun when Ziye Zhang began matriculation in Cornell's Ph.D. program in Regional Science in the fall of 2014 and continued through the completion of his doctorate and his post-doctoral research appointment at Cornell University and into his present post-doctoral research appointment at Princeton University. This work has been a most ambitious undertaking (as is discussed in Chap. 7) but has demonstrated that a dynamic commodity-flow model based on NEG behavioral foundations captures very well the stylized facts of the evolution of commodity flows. The completion of this undertaking would not have been possible without Dr. Zhang's ingenious development and deployment of Python wrappers to implement Clifford Wymer's (2006) nonlinear quasi-FIML continuous-time estimator for multiple models simultaneously on several multiple-processor workstations. This estimation work has truly been a "big data" application of computational regional science. The data needed to conduct this estimation, supplemental to the commodity-flow data derived from the REIM model, are also available from a Springer website associated with this book.

While, in setting out on this intellectual journey, we aspired to conduct dynamic game simulations with the estimated model, the enormity of such an undertaking (and the amount of time already expended on this project) mitigated against doing so. Nevertheless, by extrapolating several decades into the future commodity flows and emission intensity factors on the basis of historical trends, we have been able to develop profiles of future point-source and non-point-source emissions of four of the U.S. EPA's criteria pollutants plus VOC predicated on the assumption that globalization of the sort experienced in the last period of our sample, 1997–2007, will continue. We have also been able to develop "environmental footprints" of production and consumption based on the analysis of flows of intermediate inputs and goods delivered to final demand on the basis of this assumption (see Chap. 8).

My co-authors and I would like to thank our Cornell colleagues Oliver Gao, Peter Hess, Natalie Mahowald, Max Zhang, and our non-Cornell colleagues Genevieve Giuliano, Geoffrey Hewings, Randall Jackson, Gianfranco Piras, Jurgen Scheffran, Zhining Tao, and Clifford Wymer for their intellectual contributions and moral support. We should also like to thank the editors of Springer Verlag's Advances in Spatial Science series for their patience during the delays in delivery of this book's manuscript to the publishers, a considerable part of which was due to my being pressed into service as Interim Dean of the College of Architecture, Art, and Planning at Cornell University.

My co-authors and I are most grateful to World Scientific Publishing Company and Elsevier Ltd for permission to use previously published material in Chaps. 4 and 5. Other material previously published in Springer publications has been reused in Chaps. 2 and 6.

The research reported in Chaps. 3, 4, 5, 7, and 8 was supported by U.S. EPA grant RD-83428301-0, "Regional Infrastructure/Air Quality Planning and Global Change."

Ithaca, NY
March 2021

Kieran P. Donaghy

References

Brown-Steiner B, Chen J, Donaghy KP (2015) The evolution of freight movement and associated non-point-source emissions in the Midwest-Northeast transportation corridor of the United States, 1977-2007. In: Batabayal A, Nijkamp P (eds) The region and trade: new analytical directions. World Scientific, Singapore, pp 177–204

Brown-Steiner B, Hess P, Chen J, Donaghy KP (2016) Black carbon emissions from trucks and trains in the Midwestern and Northeastern United States from 1977 to 2007. Atmosp Environ 129:155–166

Donaghy KP (2006a) Modeling the evolution of the goods movement supply chain and associated impacts on metropolitan areas: from microfoundations to system effects. Paper presented at the National Urban Freight Conference, Long Beach, CA

Donaghy KP (2006b) Market integration, growth, and inequality. Working paper. International Programs and Studies, University of Illinois at Urbana-Champaign, Champaign, IL

Donaghy KP (2009) Modeling the economy as an evolving space of flows: methodological challenges. In Nijkamp P, Reggiani A (eds) Complexity and spatial networks: in search of simplicity. Springer, Berlin, pp 151–164

Donaghy KP (2012) Urban environmental imprints after globalization. Reg Environ Change 12:395–405
Donaghy KP (2017) Some extensions to interregional commodity-flow models. In: Shibusawa H, et al (eds) Socioeconomic environmental policies and evaluations in regional science. Springer, Berlin, pp 161–175
Donaghy KP, Chen J (2011) Generating spatial time series on interstate inter-industry freight flows. Paper presented to the National Urban Freight Conference, Long Beach, CA, October 14
Donaghy KP, Scheffran J (2006) A dynamic game analysis of network externality management. Paper presented at the 12th international symposium on dynamic games and applications, Sophia Antipolis, France, July 2006
Scheffran J, Donaghy KP, Piras G, Hewings GJD (2006) Dynamic games in complex transportation networks. Paper presented at the INFORMS 2006 meetings in Pittsburgh, PA, November 6–9
Wymer CR (2006) Systems estimation and analysis programs. WYSEA, London

Contents

Chapter 1
Introduction

Abstract This chapter presents the motivation for the research presented in this volume and provides an overview of the contributions of each chapter.

Perhaps the most significant change in the global economy over the past several decades has been the extent to which globalization—or the integration of economies at all scales—has proceeded. Over this time periods, the world has become crisscrossed with interdependent infrastructure-based networks of various sorts and the global economy has increasingly come to be seen as, in Castells' (2000) terms, a *space of flows*. Moving along the links of these networks are ever greater quantities of people, goods, material, money, and information. Settlements, in turn, appear as increasingly interdependent nodes through which these vast quantities pass.

The acceleration of flows through space can be accounted for largely by technological advances in communications and transportation and the emergence of far-flung value chains, which are driven by economizing behavior and abetted by increasingly liberal trade agreements and industrial deregulation. These developments have enabled firms to exploit *economies of scale* and *scope* by fragmenting production processes and dispersing activities to least-cost locations (Jones and Kierzkowski 2001).

Consequently, the production of most goods worldwide now takes place in a distributed pattern over many locations in which semifinished goods are shipped from one *specialized* establishment to another. *What* activities are carried out and *where* they agglomerate appear to be *path dependent*—initial advantages are reinforced due to scale effects (Venables 2006). And with the increased use of just-in-time inventory management methods, all production has become more *transport intensive*.

The obverse of this development is that a substantial amount of freight shipments are now between establishments operating in the *same* industry. As a consequence, the industrial cores of many regional economies, formerly characterized by inter-industry trade, have become *hollowed out* and regional economies worldwide have become increasingly interdependent through global supply chains (Munroe et al. 2007; Hewings and Parr 2009).

K. Donaghy et al., *The Co-evolution of Commodity Flows, Economic Geography, and Emissions*, Advances in Spatial Science,
https://doi.org/10.1007/978-3-030-78555-0_1

The growth in production and the increasing transport intensity of production have led to concomitant increases in point-source and nonpoint source emissions, but changes in production and transport technologies, induced by increasingly strict government regulations, have led to reductions in concentrations of critical pollutants in these emissions.

A broad-based community of stakeholders—politicians, planners, municipal administrators, environmental groups, port facility managers, shippers, carriers, freight handlers, and labor unions—is concerned about these developments, in large part because they lack a clear sense of how these interdependent developments are related and what they portend. Moreover, the design of effective policies to accommodate further anticipated increases in freight movement and to promote public/private partnerships that can abate and mitigate deleterious externalities—including atmospheric emissions—requires a better understanding of the coevolution of commodity flows, economic geography, and atmospheric emissions.

While empirically supported theoretical explanations of fragmentation at the firm and industry levels and network externalities at the systems level are available (e.g., Feenstra 1998; Jones and Kierzkowski 2001), we still lack theories and models that *explicitly link decision-making of producers (or shippers) and carriers to changes in commodity flows and economic geography and environmental impacts on nodes as well as links in transportation networks.*

Donaghy (2009) has elaborated an empirically oriented framework that can characterize in large the evolution of goods movement, in which the state of affairs described above can arise along with changes in emissions due to production and transportation. For such a model to be useful in forecasting exercises and thought experiments concerned with future freight movements and associated emissions and air quality, or to test theories about the evolution of freight movement and the corresponding geography of production, its parameters must be econometrically estimated. To estimate the parameters of this model, however, spatial time-series on interstate inter-industry sales must first be generated, from which inter-industry commodity flows and freight movements may be derived.

This book presents a research program whose intent is to bring the moving picture of the coevolution of commodity flows, economic geography, and atmospheric emissions into sharper focus. Whereas other studies have commented on changes in the *anatomy* of regional economies—or sectoral composition—this book focuses on the changing *physiology* of interdependent regional economies: that is, changes in how different industries function together over space and what effects they have on the environment in doing so.

Because an animating concern of this study has been to determine how changes in commodity flows and economic geography have affected atmospheric emissions and more broadly the natural environment, the next chapter is focused on the environmental impacts of globalization. The chapter provides a review of developments associated with globalization and how these developments have affected the environmental footprints of cities and states or regions in which cities are located. The chapter also discusses ways in which urban environmental footprints can be lightened. A critical part of any successful program with such an objective will involve providing new and maintaining existing urban infrastructure systems that deliver food, energy, and water to cities and facilitate interurban—hence interregional and international trade. The chapter thus concludes with a discussion of how one might provide analytical support for managing changes in urban infrastructure systems to lighten environmental footprints of cities, their hinterlands, and more distant settlements with which they trade.

Because a major impediment to the empirical analysis of the evolution of commodity flows has been the lack of spatial time series on interregional (interstate) interindustry sales that may be used to track this evolution, Chap. 3 presents and demonstrates a methodology for generating such data over the period of 1977–2007. The methodology embodies an approach to benchmarking and estimating a dynamic multiregional econometric input–output (I–O) model (or REIM) with annual data, backing out annual interregional interindustry sales coefficients from the estimated model, and using the coefficients to generate annual observations on commodity flows. The application of this methodology is demonstrated with a REIM that has been estimated with time-series data published by the US Bureau of Economic Analysis and the US Bureau of Labor Statistics for 13 industries in 13 Midwestern and Northeastern states of the United States (US) and the rest of the country. The chapter concludes with brief observations on patterns of change in same-sector interstate commodity flows and I–O sales coefficients.

Whereas Chap. 3 discusses the generation of spatial time-series data on interstate interindustry trade flows for the Midwestern and Northeastern states of the United States and the rest of the country, Chap. 4 provides an analysis and graphical portrayal of these data and detailed commentary on changes in aggregate volumes of shipments (reflecting the increased transport intensity of production and consumption), the emergence of increasing intra-industry shipments (reflecting changes in economic geography), and a preliminary commentary on changes in patterns of associated black carbon emissions that result from changes in freight movement.

Chapter 5 provides a more detailed examination of the evolution of black carbon (BC) emissions than was given at the end of the previous chapter. It presents a framework for estimating BC emissions from heavy-duty diesel vehicles (HDDV), or trucks, and trains engaged in transporting freight in the Midwestern and Northeastern United States between 1977 and 2007. The chapter also provides a comparison of estimates produced with other existing emissions inventories. The framework is then employed in attempting to answer two questions: (1) What were the trends in BC emissions from HDDV and rail transportation sources over this period and what were the major factors that drove these trends? (2) What economic sectors dominated BC emissions and what major changes in sectoral behavior occurred over this period? The framework presented also allows for the direct estimation of BC emissions under a variety of economic, technological, and regulatory scenarios through changes in transportation patterns and emission factors.

As of Chap. 5, a structural-equation model that embodies a theoretical explanation of the evolution of commodity flows, and so also attendant changes in economic geography and atmospheric emissions, has not yet been elaborated. Chapter 6 demonstrates how features of commodity-flow models within the tradition of interregional input–output modeling and the tradition of spatial-interaction modeling can be integrated into two new commodity-flow model specifications—static and dynamic. The chapter also presents extensions to extant commodity-flow models, including explicit treatment of trade in intermediate goods, so-called New Economic Geography (NEG) behavioral foundations for production, interindustry and interregional trade, and endogenous determination of capital investment and employment (Krugman and Venables 1995). It is our working hypothesis that these extensions will enable commodity-flow models to capture the features of globalization that are primary candidates as *explanans* of the developments this book addresses.

In Chap. 7, we present the operationalization and econometric estimation of a modified version of the dynamic continuous-time structural-equation model of commodity flows elaborated in the previous chapter. The objectives of this exercise are threefold: (1) to estimate an empirically based dynamic model that can accommodate the stylized facts of globalization noted earlier in this volume; (2) to determine whether or not a model that embodies an NEG formulation of production behavior is supported by the data, and (3) develop and make available for other scholars regional economic data that supplement the commodity-flow data whose derivation and analysis have been discussed in Chaps. 3–5.

Prior to this volume's publication, no one has been able to extrapolate what both point-source and nonpoint-source emissions would be for specific industries in specific locations if recent trends in globalization were to persist into the future or extrapolate what the environmental footprints (in terms of atmospheric emissions) of production and consumption in specific locations would be. Chapter 8 presents projections from 2008 to 2030 of point-source emissions of three of the US EPA's criteria pollutants—carbon monoxide (CO), nitrogen oxide (NOx), and sulfur dioxide (SO_2)—and volatile organic compounds (VOC) from industrial production for nine state or multistate groupings and nonpoint-source emissions of the criteria pollutants and VOC plus black carbon (BC) from commodity flows along stylized routes connecting centroids of these state and multistate groupings. This chapter also presents calculations of environmental (emissions) footprints by industry, pollutant, and location. This study is one of the first to make such calculations, taking into account interregional inter-industry sales that constitute commodity flows, and presents and demonstrates a methodology for doing so.

Concluding Chap. 9 summarizes the argument developed in the previous chapters.

References

Castells M (2000) The rise of the network society, 2nd edn. Blackwell, Oxford

Donaghy KP (2009) Modeling the economy as an evolving space of flows: methodological challenges. In: Reggiani A, Nijkamp P (eds) Complexity and spatial networks. Springer, Berlin, pp 151–164

Feenstra RC (1998) Integration of trade and disintegration of production in the global economy. J Econ Perspect 12:31–50

Hewings GJD, Parr JB (2009) The changing structure of trade and interdependence in a mature economy: the US Midwest. In: McCann P (ed) Technological change and mature industrial regions: firms, knowledge, and policy. Elgar, Cheltenham, pp 64–84

Jones RW, Kierzkowski H (2001) A framework for fragmentation. In: Arndt SW, Kierzkowski H (eds) Fragmentation: new production patterns in the world economy. Oxford University Press, New York, pp 17–34

Krugman P, Venables AJ (1995) Globalization and the inequality of nations. Q J Econ 110:857–880

Munroe D, Hewings GJD, Guo D (2007) The role of intra-industry trade in inter-regional trade in the Mid-West of the US. In: Cooper RJ, Donaghy KP, Hewings GJD (eds) Globalization and regional economic modeling. Springer, Heidelberg, pp 87–105

Venables AJ (2006) Shifts in economic geography and their causes. Federal Reserve Bank of Kansas City Economic Review, Fourth Quarter, pp 61–85

Chapter 2
Environmental Impacts of Globalization

Abstract This chapter reviews developments associated with globalization—among them, dramatic changes in patterns of trade and the location of economic activities—and how these developments have affected the environmental footprints of cities and states or regions in which cities are located. It also discusses ways in which urban environmental footprints can be lightened. A critical part of any successful program with such an objective involves providing new and maintaining existing urban infrastructure systems that deliver food, energy, and water to cities and facilitate interurban—hence interregional and international—trade. The chapter thus concludes with a discussion of how one might provide analytical support for managing changes in urban infrastructure systems to lighten environmental footprints of cities, their hinterlands, and more distant settlements with which they trade.

2.1 Introduction

An important observation of the early location theorist Johannes von Thünen (1826), and one understood by location theorists since Thünen's time, is that *trade* (in agricultural and other goods) and the *location* of economic activities are two sides of the same coin. (See especially Isard (1956) and Bröcker (2010).) To understand *what economic activities occur in which locations and*, and as importantly, *what impacts those activities impose on natural systems*, both regionally proximate and far flung, we need to examine *evolving patterns of trade in goods and services between firms and households within and between cities and other locations, and vice versa*. This observation suggests that the influence of cities and other hubs of economic activity on natural systems is expanding and changing with cities' growing interdependence.

Many dramatic changes in patterns of trade and the location of economic activities are associated with "globalization"—"the closer integration of countries and peoples of the world ... brought about by the enormous reduction in costs of transportation and communication, and the breakdown of barriers to the flows of goods, services, capital, knowledge, and (to a lesser extent) people across borders

K. Donaghy et al., *The Co-evolution of Commodity Flows, Economic Geography, and Emissions*, Advances in Spatial Science,
https://doi.org/10.1007/978-3-030-78555-0_2

(Stiglitz 2002, 9)." In fact, globalization has radically changed the way we conceive of cities and regions themselves (Vickerman 2007). Cities may now be viewed as *nodes* in overlapping and interdependent *networks* in a larger *space of flows* (Castells 2000; Donaghy 2009a).

With the emergence of what Manuel Castells has termed a global network society, daily activities in which urban dwellers engage affect people living half a world away and will affect generations to come. The traceability of the impacts of our actions or failures to act weighs heavily upon us and contributes to a growing cosmopolitan belief that justice, and particularly environmental justice, is owed to all, regardless of location, or other individuating features (O'Neill 2000; Elliott 2002). This belief gives urgency to the imperative to lighten our environmental footprints, wherever in the global space of flows we may be situated.

This chapter reviews developments associated with globalization and how these developments have affected the environmental footprints of cities and states or regions in which cities are located. It also discusses ways in which urban environmental footprints can be lightened. A critical part of any successful program with such an objective will involve providing new and maintaining existing urban infrastructure systems that deliver food, energy, and water to cities and facilitate interurban—hence interregional and international—trade. The chapter thus concludes with a discussion of how one might provide analytical support for managing changes in urban infrastructure systems to lighten environmental footprints of cities, their hinterlands, and more distant settlements with which they trade.

2.2 Trends Characterizing Globalization

As most analysts have pointed out, globalization—*qua* increasing global economic integration—is not new. There have been significant periods of globalization over many centuries. What *is* new is the increasing speed of the movement of goods and services, people, capital, and technology around the world and the opportunities and challenges this *velocitization of flows* presents. Five trends that characterize the most recent phase of globalization are (1) an accelerated reduction in transport and communication costs, (2) greater international specialization driven by trade liberalization, (3) increasing trade in services due to the revolution in information and communications technologies, (4) increasing integration of emerging markets into the world economy, and (5) the consolidation of production systems. In taking notice of these trends, we do not overlook the fact that, as Ghemawat (2007) has observed, significant differences between countries—in culture and business practices, *inter alia*—continue to serve as impediments to closer economic integration.

Technological advances in transportation have reduced costs of transport while increasing speed and improving its reliability. As evidence in support of this point, consider that between 1920 and 1990 the average charge per ton in constant prices for ocean freight and use of port facilities decreased by about 70% (HM Treasury 2004). Such advances and the refinement of *logistics*, or the scientific management of the flow of materials through an organization from raw materials to finished products, have extended the geographical reach of firms by making new markets accessible on a cost-effective basis (Economist 2002). Advances in information and communications technologies (ICT) have facilitated the growth of global transactions and improved information flows, thereby increasing productivity in almost all areas of economic activity. The extent of the rate at which advances in ICT have proceeded is dramatically illustrated by the fall in the cost of one megahertz of processing power between 1970 and 1999 from $7601 to 17 cents (HM Treasury 2004).

In his book, *The World is Flat*, Thomas Friedman (2005) made much of the role played by ICT, and this role should not be underestimated. As Russel Cooper and Gary Madden (2004) have argued, ICT is not just another *leading sector*, such as aviation, which is based on an innovation that revolutionizes only a small part of the economy. ICT has contributed to growth throughout the economy via linkages between ICT-embedded products and positive network externalities, or *spillovers*, between industries. In the form of e-commerce, it has changed the very way that business is conducted. ICT investment in developed countries has been a driver of growth because it has been growing faster than labor as an input, hence it contributes to average labor productivity via *capital deepening*. In less developed countries, a similar positive relationship between ICT investment, growth, and productivity has been slower to develop. This delay would appear to be due to low ratios of ICT to gross domestic product and a lack of complementary assets, in the form of network infrastructures and knowledge bases, to support effective use of ICT. For these reasons, the ICT revolution is increasing the North-South "digital divide." As will be discussed below, there are also digital divides *within* developed and developing countries.

Technological change has also affected the structure of firms and the location of ownership and management of productive activity among regions and countries. A new *functional division of labor in space* has emerged enabling firms to allocate operations according to comparative advantages within and between countries. As a result, there has been a significant increase in the number of firms that locate, source, and sell interregionally and internationally, reflecting the new opportunities presented by the ICT revolution, alongside falling transport costs and easing of trade and capital restrictions (Marsh 2005). The worldwide web and the Internet have also changed the way many markets operate. Such technologies have contributed to further integration of interregional and international markets, enabling more effective competition and international convergence in prices for traded goods and services. They have also facilitated fundamental changes in the way businesses are run. The combination of technological advance and trade liberalization has contributed to the increased specialization, internationalization, and dispersion of production processes. An important aspect of this internationalization and subdivision of activities within and between firms is the increasing relocation of economic activity abroad—international *outsourcing* or *offshoring*. (See Edwards (2004).)

Outsourcing has presented huge opportunities for countries with developed and developing economies alike. While not new, it has been used more effectively than in the past because technological advances have lowered the costs and enhanced the ability of firms to manage complex, geographically dispersed, and highly specialized production processes. For example, innovations in logistics have enabled firms to source components internationally. In 2005, about 80% of the parts of an automobile assembled in the United States were manufactured abroad (Uchitelle 2005). Policy liberalization has permitted cross-border activities that were previously prohibited and lowered costs of others—e.g., insurance, banking, and transportation. Developments in foreign markets have made economically viable the foreign provision of certain goods and services that was previously not so.

Certainly, the specialization and internationalization of activities formerly domiciled in domestic economies have led to an increasing trade in ICT-enabled services as well as semifinished goods. This increasing trade has permitted outsourcers to enjoy cost advantages but has also helped developing countries to move up the value chain, or engage in activities contributing more value-added to final products (Baldwin 2016). It is important to understand that the dramatic increase in international trade is in large part due to the increased trade in services and semifinished goods and that this increase coincides with international industrial restructuring. In many cases, the jobs being created abroad are not the ones being lost at home. Because outsourcing presents an opportunity to make a change in the nature of the production process itself, not just the location of production, the lost jobs may never reappear *anywhere*. Moreover, many countries that outsource activities in the production chain (or import services), such as the United Kingdom and the United States, are themselves net *exporters* of services—the increasing trade in services flows both ways between developed and developing countries.

The growth of trade in services has been mirrored in a shift in the composition of global foreign direct investment (FDI) and given developing countries an expanding share of world trade. The rapid expansion of emerging economies is leading to a shift in the global balance of economic activity, to shifts in shares of world trade, and increased demand in the world's commodity markets. Whereas in previous decades the United Kingdom and United States and other "safe havens" were primary destinations of FDI, China and the countries of Africa have become the largest recipients in recent years. One implication of increased FDI in places such as China and Africa has been increased environmental pollution from the industrial activity begotten at the site of the capital investment.

The final trend characterizing globalization to be discussed here is that of consolidation of production systems. In the late 1990s and early 2000s in particular, there was a steady progression of mergers and acquisitions, occurring under various circumstances for various reasons, which have enabled dominant companies to reconfigure global markets (Krugman 2002; Lynn 2005). Lynn observes, "Rarely challenged by government, ... deals have led to instances of consolidated power we have not witnessed in nearly a century. Consider that Owens-Illinois now produces more than half the world's food containers. Or that Intel supplies 90% of the world's demand for certain key semi-conductors (Lynn 2005)." To this, we might add that Archer Daniels Midland and Cargill controlled [at that time] major shares of agricultural commodities markets from the point of purchase of raw materials to penultimate stages of processing. These observations are not made to impugn the actors involved for their size or scope of operations, only to point out how dependent in an increasingly interdependent world other industries, and the communities in which they are based, have become on a limited number of central actors.

2.3 Effects of Globalization

It has been observed by Friedman and others that consumers in both advanced and emerging economies have benefited from the increase in the range of goods and services that greater global integration has made possible. Indeed, goods and services *have* become cheaper as global supply chains have been reorganized to make best use of resources that emerging economies have to offer. Increased integration has also expanded the range of assets available to investors, providing greater scope to diversify their investment risks. And, on balance, as a consequence of globalization, the number of people and proportion of the world's people living in extreme poverty has fallen, while the "welfare of humanity, judged by life expectancies, infant mortality, literacy, hunger, fertility, and the incidence of child labour has improved considerably (Wolf 2004, 171)."

However, globalization also has its *dis*-benefits. For one, the entire system of international commerce has become increasingly dependent upon—and vulnerable to disruptions in—interdependent infrastructure-based networked systems. (See Donaghy et al. (2005b) and coverage of supply-chain disruptions caused by the earthquake and tsunami at Fukushima in Japan (e.g., Marsh (2011).) Moreover, changes in the nature and location of economic activities have had significant negative impacts on particular groups of individuals, activities, and areas. This phenomenon is not surprising. As economic growth theorists from Joseph Schumpeter (1934) onwards have pointed out, development is by its very nature uneven or unbalanced and often occurs through a process of "creative destruction," in which new industries supplant old ones and whole classes of occupations disappear as new ones emerge.

While the process of globalization described above has indeed promoted a leveling of per capita incomes between countries, it has also contributed to greater disparity in income per capita between social classes within countries and between regions of countries, both developed and developing. (The *Financial Times* has reported that there is also "greater variation in economic growth rates between regions within major countries than between countries themselves (Briscoe 2005, 16).") Martin Wolf has observed that "inequality has risen in most high-income countries. Where it has not risen, unemployment has tended to increase instead (Wolf 2004, 168)." This inequality is due to widening gaps in relative pay between skilled and unskilled workers. Factors contributing to such gaps include rising imports of labor-intensive manufactured goods, institutional changes that have weakened the bargaining position of labor unions, technological changes, immigration of unskilled labor, and insufficient supply of skilled labor—a domestic version of the "digital divide" alluded to earlier. The so-called "skill gap" in pay and indeed in employment has contributed to poverty-driven environmental degradation *within* metropolitan areas of developed countries. (Witness the degradation of Detroit and Cleveland, old industrial centers in the United States, in the wake of the 2008 financial crisis.)

The trends discussed above account for much of the transformation that has been experienced by many regional economies. Declining costs of transportation and communication and the fragmentation of production activities have allowed firms to exploit *economies of scale* that specialization allows. Consolidation of production systems has also permitted semifinished goods and services to be used in the intermediate stages of production across different product lines, thereby permitting the realization of *economies of scope.* (See Jones and Kierzkowski (2001).) A consequence of these developments—and one that will concern us through much of the rest of this book—is that production processes have become increasingly *transport intensive.* With increasing amounts of trade (through outsourcing) occurring within the same industries and most of the interactions between establishments of firms now occurring at greater distances, multiplier effects in local and regional economies have diminished and the industrial bases of these economies have become "hollowed out." Now, when expansion or contraction occurs at a branch plant, the largest impacts are likely to be experienced at a more centrally placed source node in the production network. Such developments constitute *evolutions in economic geography* to which the title of this volume alludes.

In countries with developing economies, the trends discussed above have often contributed to the driving out of indigenous industries and local sourcing networks, as these countries become sites of unskilled assembly operations. Unless localities in such countries can move quickly up the value chain of production and establish backward linkages to their domestic industries, they stand to lose newly gained branch-plant operations to other developing economies along with their indigenous industries. (See, however, Thompson's (2005) reporting on the successful up-leveling of skills and value contribution by *maquilladora* operations in Mexico.) Of perhaps greater concern is the off-loading of deleterious environmental externalities to remote production sites associated with lower levels of the value chain.

The transport intensiveness of production has placed great stress on not only the transport system. As just-in-time production and inventory management systems have helped displace assembly lines effectively from factory floors to sequences of intersections on interstate (and international) highway systems (or river- and railways), key nodes in transport networks (especially transshipment points, such as ports) have come to bear the negative externalities of congestion, pollution, and changes in land-use patterns. Neither this situation nor current rates of energy consumption by transport are sustainable. (See Donaghy et al. (2005a).)

The OECD has identified a number of effects of globalization on the size of a national economy's "ecological footprint," which are associated with the effects discussed above (OECD 2001a). These effects include *scale effects*, or effects from the acceleration of economic growth as a result of increased rationality of economic processes following from a country's broad participation in the international division

Table 2.1 Environmental effects of globalization: positive and negative aspects

Effects	Positive aspects	Negative aspects
1. Scale effects	Reduction of poverty-driven degradation Growing ecological consciousness	Increased use of resources and energy Increased waste production
2. Structural effects	Decreased use of energy and resources due to better allocation of production	Losses caused by mistakes in pricing of ecological assets, shifting activities to damaging sectors
3. Technology effects	Decreased use of energy and waste production due to technology transfer	Technology dumping
4. Product effects	Increased consumption of environmentally *less* damaging goods	Increased consumption of environmentally *more* damaging goods Increased transport intensity
5. Regulatory effects	Positive environmental impact of international trade agreements	Negative environmental impact of international trade agreement

Sources: OECD (2001a) and Budnikowski (2006)

of labor; *structural effects*, or effects from shifts in the structure of production internally and internationally; *technological effects*, or effects from accelerated diffusion of technology as a result of liberalization of international trade and capital flows; *product effects* or effects from liberalization of economic relations with foreign countries resulting in changes in the mix of manufactured and consumer goods; and *regulatory effects*, or effects from the adoption of international trade agreements.

Given that the vast majority of economic activity in the world's largest economies is based in cities (Glaeser 2011; McCann 2011), we can view the OECD's notion of a nation's ecological footprint as a spatial generalization of the concept of an urban environmental footprint, which we have been employing. Positive and negative aspects of these effects identified by the OECD are summarized in Table 2.1.

Quite clearly in twenty-first century supply chains, energy sources and foodstuffs are being transported over ever-greater distances, while, with some notable exceptions, water remains a more locally provided resource. (The city state of Singapore, however, imports over half of its fresh water (International Business Machines Corp 2009).) Given the changing (or changed) environmental footprints of cities, it seems reasonable to ask: What can be done to lessen the imposition of cities' negative environmental externalities—or, more generally, those of networks of production and consumption—near and far?

2.4 Mitigating the Environmental Impacts of Globalization

As urban populations continue to grow and ever-increasing demands are placed upon sources of food, energy, and water, a perpetual expansion of the environmental footprints of cities is not fated to be. In evidence to the contrary, the notion of "eco-cities" or "smart cities" is increasingly taking hold of urban planners and engineers. Eco-cities can be defined as "areas where urban planning and environmental management tools are applied to pursue synergies in resource utilization and productivity, waste management, environmental preservation, industrial and economic development and a healthy living environment (Eco-Cities Forum 2011)." Smart cities are urban areas in which ICT-embedded technologies are implemented to monitor and respond to conditions so as to maximize the efficient use of scarce resources.

Efforts to build eco-cities and smart cities or to transition to them are increasing in Southeast Asia and the Middle East in particular, both of which are water-scarce areas (Biello 2008). IBM's Global Innovation Outlook (GIO) Report on Water observes that "alongside solar power, wind turbines and energy efficient buildings, ... smart cities will include dew catchers, rain water harvesting, low-energy desalination, and electric sensors to detect leaky pipes. Gray water will be used to water urban crops, grown in vertically stacked high-rise plots, and the water not used by plants will be recovered and re-used (International Business Machines Corp 2009, 32)."

In the energy sector, electrical engineers are developing "smart" power grids to link distributed generation by electricity producers at power stations and wind and solar farms with consumers in offices, factories, and homes (Crooks 2009). In this sector, electric or hybrid vehicles will have a vital role to play in storing power and balancing fluctuations (Cookson 2010).

Not only new cities are adopting approaches to lighten environmental footprints. Stockholm, selected in 2010 as the first Green Capital of Europe, has increased use of public transit by its populace and exploited efficiencies of district heating and waste-to-energy treatment by undertaking long-term planning with environmental impacts in mind and working with its citizens to build consensus for more sustainable lifestyles (Partnership for New York City 2010).

While isolated efficiency improvements in the production or usage of food, water, and energy are encouraging, it is important to recognize that changes in production or use of these resources are interrelated. Energy is needed to produce food and food can be used to produce energy. Energy is needed to clean and transport water, but water is needed to generate energy. Water is needed to grow food, whereas food transports (virtual) water (International Business Machines Corp 2009). Given these fundamental relationships, a broad perspective of integrated resource management needs to be adopted. According to IBM's GIO Report, this perspective requires a "global approach to managing the planet's food, energy, water and even climate. Make a local change to one of these systems and consequences will be felt throughout the global chain. For example, freshwater can be produced from desalination, but it consumes a great deal of energy, which negatively affects the climate. And energy can be produced from bio-fuels, but [such fuel production] can strain global food supplies and requires large quantities of water, and so on (International Business Machines Corp 2009, 41)."

In its *Environmental Outlook 2001*, the OECD recommended national policy packages that combine policy instruments to target the range of actors affecting the environment, draw on synergies for realizing different objectives while avoiding policy conflicts, and address social or competitiveness concerns about the policy instruments (OECD 2001a). The OECD has also observed that implementation is the weak link in the environmental policy cycle. It notes that successful implementation will require setting objectives and targets, using indicators, collecting data, involving stakeholders, and adopting effective compliance and enforcement mechanisms (OECD 2001b, 20).

At the international level, the European Union, whose very organizational principles—promoting the free movement across borders of goods, people, and capital—are the embodiment of globalization, has recognized the need to take into account the environmental effects of the increasing transit-intensity of modern globalized life and has insisted that aviation and shipping sectors, which enable the essential flows between the cities of different nations, be included in global carbon-cutting measures in any future climate accords (Pignal et al. 2009).

Different steps can also be taken by *nongovernmental* actors. Private sector firms are adopting measures that mitigate climate change and other environmental impacts, recognizing that not only do such measures reduce operating costs along with impacts but that it is in their interest to face a common set of environmental standards that are followed by all competitors in their industries (Duncan 2007). While global supply chains are not likely to disappear soon (Hill 2011), large firms such as Wal-Mart, John Deere, and Caterpillar have begun "greening" their supply chains by entering into agreements that bind firms they purchase from to the same environmental standards of production that they observe (Donaghy 2007). Given the consolidation of production systems and the lengthening of supply chains, remarked upon above, this development is of great significance.

Even as the number of megacities, or cities with populations larger than ten million, around the work is increasing annually, we should also note that other cities, especially ones that have lost their industrial base or which are in countries with low birthrates, are losing population and decreasing their physical footprint. Although of smaller size, these cities' residents still impose environmental impacts locally and at great distances to the extent they are integrated into a globalized world. While planned downsizing may be appropriate in some cases, to truly reduce a city's environmental footprint the questions of aggregate and per capita levels of consumption, efficiency of resource use, and local sourcing of life's necessities must also be revisited.

Realism compels us to acknowledge that as the size of the global population approaches eight billion and continues to increase, and as projections of the share of this population that resides in urban areas climb higher and higher, it will be difficult to mitigate the environmental impacts of urban life through government policies, interfirm agreements, and/or technological efficiency gains unless there are also lifestyle changes at the household level that result in absolute reductions in consumption (Crocker and Linden 1998; Arrow et al. 2004).

Aside from facilitating dramatic lifestyle changes, perhaps the single most important action that can be undertaken to reduce the environmental imprint of cities in a world of increasing economic integration is to invest in the construction and

Table 2.2 Projected infrastructure by region and sector in $US trillion, 2005–2030

	North America	Latin America	Europe	Africa	Middle East	Asia Pacific	Total
Water	3.62	4.97	4.52	0.23	0.23	9.04	22.61
Electricity	1.53	1.44	1.08	0.54	0.18	4.23	9.00
Road and Rail	0.94	1.01	3.12	0.31	0.31	2.11	7.80
Air and Seaport	0.43	0.06	0.43	0.02	0.14	0.51	1.59
Total	6.52	7.48	9.15	1.10	0.86	15.89	41.00

Source: Cohen and Steers Global Infrastructure Report 2009: the $40 Trillion Challenge

maintenance of efficient urban infrastructure systems. At this time, much of the urban infrastructure built within the last century has reached the end of its service life. Consider the case of the United States. The American Society of Civil Engineers (ASCE) estimates that nearly 30% of the bridges in the United States are structurally deficient or functionally obsolete (ASCE 2009). The number of unsafe dams has increased by nearly 33%, and funding for public transit facilities, drinking water, and wastewater management is deemed to be grossly inadequate. A similar state of affairs confronts many European and Asian countries. (See Estache (2004), Jones (2006), Katz et al. (2009), and Timmins (2010).)

IBM's GIO Report on Water observes that, because it is expensive, difficult to maintain, and politically unpalatable, infrastructure is the water-related imperative that is most easily ignored. The critical need for investment in water infrastructure is borne out by the fact that, in some cities, 15–20% of available water is lost to leaks and in developing countries it is worse (International Business Machines Corp 2009, 31).

According to Cohen & Steers' *Global Infrastructure Report* for 2009, the world is poised (or needs) to undertake $40 trillion in both new and sustaining infrastructural investment over the 25-year period of 2005–2030, most of which will be in or around cities. (Please see Table 2.2.)

2.5 Managing Changes in Urban Infrastructure Systems

Urban and regional planning is concerned with managing changes in territorially organized systems (Friedmann 1987). Managing changes in infrastructure systems is particularly challenging because, in repairing old or implementing new infrastructure systems, planners are simultaneously endeavoring to bring about changes in behavior supporting more sustainable lifestyles. Because, as noted above, production or usage of food, energy, and water are interdependent, the urban infrastructure systems on which their delivery is based are also interdependent. Hence, the management of changes in urban infrastructure systems must take this interdependence into account.

As in most planning problems, managing changes in urban infrastructure systems to lighten urban environmental footprints will entail determining—not just once, but in a recurring fashion—*what is to be done by whom, when* and *where,* and *by how much.* Each element of this determination is critical to an appropriate planning response. (See Donaghy and Schintler (1998).) Of course, given the complexity of the relationships between interdependent infrastructure-based networked systems, their jurisdictions, their controllers, and the nature of their financing, it is clear that there will be multiple agents—both public and private—engaging with this planning problem, that their decisions will affect each other, and that the overall infrastructure system will be a complex adaptive one that is likely to give rise to emergent outcomes that no single "network controller" will have intended. (See Donaghy (2009a), Torrance (2009), and Donaghy (2011).) Perhaps the best we can hope to accomplish is to identify plans that are compatible within a web of overlapping plans of agents and planning jurisdictions. (See Donaghy and Hopkins (2006).)

Still, and in spite of this inherent indeterminacy, we will need to employ modeling tools to identify possibly compatible systems management plans and so we will need to know what theoretical and methodological resources are available. When we survey the availability of theoretical and methodological resources to support the management of change in urban infrastructure systems to lighten urban environmental imprints, we may wish to avail ourselves of contributions from systems engineering, operations research, network science, and game theory, as well as urban and environmental economics. Network science, which identifies and describes recurring self-organizing behaviors in networks, is particularly appropriate because its insights are critical to designing intervention schemes—indicating who is to do what, when and where, and by how much—appropriate to a network controller's objectives (Barbasi 2002).

Friesz et al. (2007) suggest that it is helpful to view infrastructure systems involved with the movement of goods (including food), passengers, information, energy, and water as *general transportation networks.* Moreover, they argue that, to the extent such networked systems are interdependent, they should be viewed together as a *system of systems* (Sheffi 1985; Nagurney and Dong 2002). They identify the five main sources of interdependence between generalized transportation networks as being (1) physical interdependence, (2) budgetary interdependence (when public financing is involved), (3) market interdependence and spatial economic competition, (4) informational interdependence, and (5) environmental and congestion externalities—a primary concern of this chapter.

The practical challenge of implementing a system of systems framework is to express the interdependencies between the infrastructure networks mathematically "so that richer and more informative models to support infrastructure network planning and design may be formulated and numerically solved (Friesz et al. 2007, p. 56)." One way to proceed is to represent infrastructure systems as multilayered networks with constraints upon how the layers are coupled. The layers can then be arranged in hierarchies reflecting their engineering and societal functions and the resulting multilayered coupling of infrastructure networks will constitute a system of systems. (See Fig. 2.1 for an illustration and Friesz et al. (2007) for a

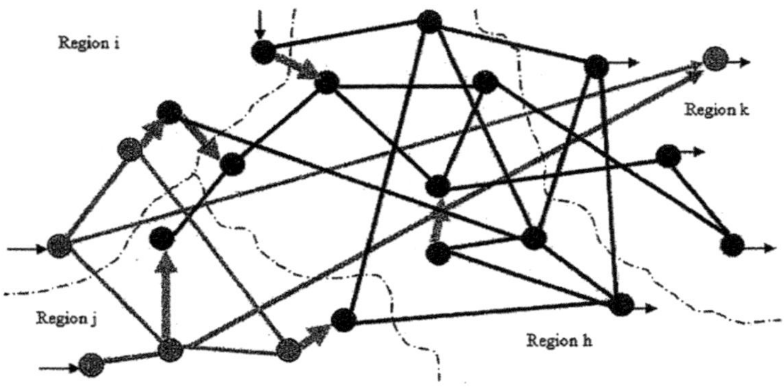

Fig. 2.1 Interdependent networks involving different sets of nodes (Donaghy et al. 2005a, b)

general algebraic representation of a system of interdependent infrastructure networks.)

Friesz et al. remark that the performance of a system of systems "can be significantly influenced by decisions taken by individuals or groups at various levels in the subsystems (p. 59)." Acknowledging this eventuality, modelers usually adopt a noncooperative game-theoretic approach to representing interdependent strategic decision-making behavior. Zhang et al. (2005) have investigated multilevel network games that correspond to systems of systems. In such games, decision makers associated with distinct tiers or subsystems may compete with other decision makers in their own tier and cooperate with other decision makers on tiers not their own.

If the layers of the system of infrastructure networks are viewed collectively as the means by which agents in a market economy complete their transactions, the modeling framework may be viewed as a spatial computable general equilibrium (or SCGE) model (Donaghy 2009b).[1] In such a model, the generalized transportation

[1]Computable General Equilibrium (CGE) modeling is an approach to applied economic analysis in which theories of economy-wide market behavior are used to impose structure in numerical thought experiments concerning matters of trade and development—and related policies—where the relative unavailability of data or the complexity of a theoretical model's specification poses problems for a more traditional analytical or econometric modeling approach. Over the past thirty years, CGE modeling has developed extensively and has become a stock in trade of regional economists in particular. More recently, CGE models have taken on an explicit spatial orientation as the focus of modeling exercises has turned to analysis of location-specific impacts of unplanned events and planned industrial, infrastructural, environmental, or other types of regional policies. Spatial CGE models have been employed by researchers at various scales of spatial and temporal resolution to examine a wide variety of phenomena. Owing to the paucity of spatial time series, spatial CGE models provide logical frameworks within which a broad spectrum of spatial economic issues may be analyzed. See Donaghy (2009b) for a fuller discussion. There are a number of problems associated with the solution of such models. First is its sheer size: such models could easily have thousands of equations. While such a dimension is not uncommon for CGE models, the models would also have path variables and unavoidable non-convexities due to the coupling of various network layers. Consequently, nontraditional numerical solutions methods must be explored,

networks can be represented in fine enough detail to support planning and engineering analyses. Such an articulation should also enable one to study the influence of specific infrastructure network features on all economic sectors, and in turn the physical environment, at all represented locations through conventional comparative statics methods. (For examples of comparative statics analyses of changes or disruptions to interdependent infrastructure-based networks, see Kim et al. (2002), Sohn et al. (2003), and Ham et al. (2005).)

The equilibria computed with an SCGE model could in turn be used to construct a dynamic model of coupled infrastructure networks based on principles of disequilibrium adjustment. Friesz et al. note that such a model could allow study of "nonlinear synergies and catastrophes among infrastructure technologies that would go unnoticed so long as the traditional one-network-at-a-time paradigm is employed (p. 60)."

SCGE models can be employed to compute an equilibrium state between spatially distinct markets in terms of steady-state flows along infrastructure networks. But for planning purposes, we would also need to be able to examine time-varying flows and other transient phenomena that are critical to the success of infrastructure and network engineering projects. Hence, we need to provide an explicit formulation of adjustment dynamics in a more encompassing modeling framework whose steady-state solutions are equilibria characterized by SCGE models.

If models portraying dynamics of interdependent infrastructure-based networked systems are to support life-cycle management of change in such systems—and hence the allocation of resources for their construction, operation, maintenance, and replacement—one must articulate allocative criteria. Friesz et al. suggest using the net present value of benefits. Other criteria might include accessibility, diversity, sustainability, resilience, maximum entropy, or perhaps some synthetic index of environmental impact. (See Ulanowicz et al. (2009).)

With allocative criteria (or a weighted combination thereof) chosen, one can proceed to the formulation of a dynamic multilayered urban infrastructure-based networked systems management model. One such model could take the form of a *capital budgeting* model whose solution would indicate the optimal effective capacity enhancement trajectories for the arcs of the infrastructure networks and the time paths of the network flows and associated costs. The resulting management plan would be conditioned on the acknowledgment that capacity perturbations give rise to disequilibria which in turn induce equilibrating adjustments. Such a model naturally embodies an inter-temporal optimization (or optimal control) problem in which the criterion function, e.g., the present value of net benefits, is maximized subject to state dynamics (or equations of motion), budget constraints, layer-to-layer coupling constraints, non-negativity constraints on network flows, and upper-bound constraints on arc capacities.

---

include variational inequality methods, simulated annealing, genetic algorithms, and agent-based modeling methods. (See Friesz et al. (2007) and Zhang et al. (2005).)

There are a number of problems associated with the solution of such a model. First is its sheer size; the model could easily have thousands of equations. While such a dimension is not uncommon for CGE models, the model would also have explicit path variables and unavoidable non-convexities, due to the coupling of various network layers. Consequently, nontraditional numerical solution methods must be explored, including variational inequality methods, simulated annealing, genetic algorithms, and agent-based modeling methods. (See Friesz et al. (2007) and Zhang et al. (2005).)

Solutions to such models will be very detailed and will need to be displayed with appropriate visualization and other decision and planning support tools to help stakeholders appreciate the economy wide and spatial implications of a given infrastructure systems management plan.

An example of a system of systems problem involving interdependent networks and strategic behavior might arise in the case of simultaneous construction of a high-speed rail system linking several cities and modification of a regional power grid to accommodate both high-speed rail and regional use of plug-in hybrid vehicles (buses, trucks, and automobiles). These changes, which of themselves could substantially affect circulation and land use patterns within cities and hence the cities' environmental imprints, might also induce significant changes in the location of industrial activities, resource usage, and patterns of freight movement throughout the region and beyond. Given that there would be both public and private sector actors from multiple jurisdictions and industries involved in these developments, coming to terms with the spatial and temporal implications of new transportation and power systems and managing the necessary changes at multiple spatial and administrative levels might be substantially aided by developing and implementing one or more modeling frameworks along the lines here discussed.

2.6 Conclusions

While cities continue to cast a large environmental footprint on their regionally proximate physical environments, they are also exerting a stronger influence on the natural systems of more remote locations because of their growing interconnectedness and interdependence with other cities—that is, because of globalization. In the foregoing, we have reviewed some of the most salient features of globalization and its environmental impacts. We have also surveyed possible responses to these developments and focused in particular on the critical role that urban infrastructure systems can play. Finally, we have discussed how one might provide analytical support for managing changes in urban infrastructure systems to lighten urban environmental imprints.

In the subsequent chapters of this volume, our analysis moves to a higher level of geospatial abstraction and a narrower interregional focus as we concern ourselves more with commerce between industry aggregates in a limited number of states and groupings of states, although through a stylized network connecting cities as centroids in states and state groupings. The empirical commodity flow data, whose generation is discussed in Chap. 2 and analyzed in Chaps. 3, 4, 6, and 7 reflect the trends of globalization discussed above—the velocitization of network flows, the increasing transport intensity of production and consumption, and the hollowing out of local economies (effecting changes in economic geography) that coincides with increasing interindustry trade and nonpoint source emissions associated with that trade. This volume's penultimate chapter portrays future environmental footprints of production and consumption if these trends should persist.

References

American Society of Civil Engineers (2009) 2009 Report card for America's infrastructure. ASCE, Reston, VA

Arrow K, Dasgupta P, Goulder L, Daily G, Ehrlich P, Heal G, Levin S, Mäler K-G, Schneider S, Starrett D, Walker B (2004) Are we consuming too much? J Econ Perspect 18:147–172

Baldwin R (2016) The great convergence: information technology and the new globalization. The Belknap Press of Harvard University Press, Cambridge, MA

Barbasi AL (2002) Linked: the new science of networks. Perseus, Cambridge, MA

Biello D (2008) Eco-cities: urban planning for the future. Scientific American, September 24. http://www.scientificamerican.com/article.cfm?id=eco-cities-urban-planning

Briscoe S (2005) Numbers in the news: the growth gap within countries. Financial Times, October 6, 2005, 16

Bröcker J (2010) Is there a German school of location theory before 1945? Manuscript

Budnikowski A (2006) Globalization and the ecological footprint. Case study: Poland. Paper presented at a conference on Economic Globalization and Environmental Policy at the Warsaw School of Economics, May 25–26

Castells M (2000) The rise of the network society, 2nd edn. Blackwell, Oxford

Cohen & Steers (2009) Cohen & Steers global infrastructure report 2009: the $40 trillion challenge. Cohen & Steers, New York

Cookson C (2010) Electric cars eyed as store of power for grid. Financial Times, February 20, 2

Cooper RJ, Madden G (2004) Understanding the evolving digital divide. In: Cooper RJ, Madden G (eds) Frontiers of Broadband, electronic, and mobile commerce. Contributions to Economics Series. Physica-Verlage, Heidelberg, pp 315–331

Crocker DA, Linden T (eds) (1998) Ethics of consumption: the good life, justice, and global stewardship. Rowman & Littlefield Publishers, Lanham

Crooks E (2009) Industry looks to green electric future. Financial Times, December 10, 6

Donaghy KP (2007) Climate change and planning: responding to the challenge. Town Plan Rev 78: i–ix

Donaghy KP (2009a) Modeling the economy as an evolving space of flows: methodological challenges. In: Reggiani A, Nijkamp P (eds) Complexity and spatial networks: in search of simplicity. Springer, Berlin, pp 151–164

Donaghy KP (2009b) CGE modeling in space. In: Capello R, Nijkamp P (eds) Regional dynamics and growth: advances in regional economics. Edward Elgar, Cheltenham, pp 731–799

Donaghy KP (2011) Managing change in urban infrastructure systems. In: Brooks N, Donaghy KP, Knaap GJ (eds) The Oxford handbook of urban economics and planning. Oxford University Press, New York, pp 81–97

Donaghy KP, Hopkins LD (2006) A coherentist approach to planning theories is possible and useful. Plan Theory 5:173–202

Donaghy KP, Schintler LA (1998) Managing congestion, pollution, and infrastructure in a dynamic transportation network model. Transp Res D 3:59–80
Donaghy KP, Poppelreuter S, Rudinger G (eds) (2005a) Social dimensions of sustainable transport: transatlantic perspectives. Ashgate, Aldershot
Donaghy KP, Vial JF, Hewings GJD, Balta N (2005b) A sketch and simulation of an integrated modeling framework for the study of interdependent infrastructure-based networked systems. In: Reggiani A, Schintler L (eds) Methods and models in transport and tele-communications: cross-Atlantic perspectives. Springer, Berlin, pp 93–117
Duncan E (2007) Cleaning up: a special report on business and climate change. The Economist 383.8531 (2 June), 1–30
Eco-Cities Forum (2011) Definition of eco-cities. http://www.eco-cities.net/Static/main.htm. Accessed May 23, 2011
Edwards B (2004) A world of work: a survey of outsourcing. The Economist, November 13, 1–20
Elliott L (2002) Global environmental (in)equality and the cosmopolitan project. CSGR working paper No. 95/02, Centre for the Study of Globalization and Regionalization, University of Warwick, Coventry, UK. http://www.csgr.org
Estache A (2004) Emerging infrastructure policy issues in developing countries: a survey of the recent economic literature. Background paper for the October 2004 Berlin meeting of the POVNET Infrastructure Working Group
Friedman TL (2005) The world is flat: a brief history of the twenty-first century. Farrar, Straus and Giroux, New York
Friedmann J (1987) Planning in the public domain: from knowledge to action. Princeton University Press, Princeton
Friesz TL, Mookherjee R, Peeta S (2007) Modeling large-scale and complex infrastructure systems as computable games. In: Friesz T (ed) Network science, nonlinear science and infrastructure systems. Springer, Berlin, pp 53–75
Ghemawat P (2007) Redefining global strategy: crossing borders in a world where differences still matter. Harvard Business School Press, Cambridge, MA
Glaeser E (2011) The triumph of the city: how our greatest invention makes us richer, smarter, greener, healthier, and happier. Penguin, New York
Ham H, Kim TJ, Boyce D (2005) Implementation and estimation of a combined model of interregional multimodal commodity shipments and transportation network flows. Transp Res B 39:65–79
Hill A (2011) 'Just in time' is not past its sell-by date. Financial Times, March 22, 10
HM Treasury (2004) Long-term global economic challenges and opportunities for the UK, October
International Business Machines Corp (2009) Water: a global innovation outlook report. International Business Machines Corporation, Armonk, NY
Isard W (1956) Location and space economy. Cambridge, MA, MIT Press
Jones S (2006) Infrastructure challenges in East and South Asia. Report prepared for the Institute of Development Studies
Jones RW, Kierzkowski H (2001) Horizontal aspects of vertical fragmentation. In: Cheng L, Kierzkowski H (eds) Global production and trade in East Asia. Kluwer Academic, Boston, pp 33–51
Katz B, Puentes R, Geissler C (2009) America's infrastructure: ramping up or crashing down? Brookings, Washington, DC
Kim TJ, Ham H, Boyce DE (2002) Economic impacts of transportation network changes: implementation of a combined transportation network and input-output model. Pap Reg Sci 81:223–246
Krugman P (2002) For richer: how the permissive capitalism of the boom destroyed American equality. The New York Times Magazine, October 20, 62–77, 141–142
Lynn B (2005) The fragility that threatens the world's industrial systems. Financial Times, October 18, 17
Marsh P (2005) No fixed abode for the modern manufacturer. Financial Times, August 17, 7

Marsh P (2011) High and dry. Financial Times, April 13
McCann P (2011) The Role of industrial clustering and increasing returns to scale in economic development and urban growth. In: Brooks N, Donaghy K, Knaap G (eds) The Oxford handbook of urban economics and planning. Oxford University Press, Oxford
Nagurney A, Dong J (2002) Supernetworks. Edward Elgar, Cheltenham
O'Neill O (2000) Bounded and cosmopolitan justice. Rev Int Stud 26:45–60
Organization of Economic Cooperation and Development (2001a) OECD environmental outlook 2001. OECD, Paris
Organization of Economic Cooperation and Development (2001b) Highlights of the OECD environmental outlook 2001. OECD, Paris
Partnership for New York City (2010) Cities of opportunity. Price, Waterhouse, Coopers, LLP, New York
Pignal S, Chaffin J, Barber T (2009) EU wants ships, aircraft in climate pact. Financial Times, October 22, 3
Schumpeter JA (1934) The theory of economic development. Oxford University Press, Oxford
Sheffi Y (1985) Urban transportation networks: equilibrium analysis with mathematical programming methods. Prentice Hall, Englewood Cliffs
Sohn J, Kim TJ, Hewings GJD, Lee JS, Jang SG (2003) Retrofit priority of transport network links under an earthquake. J Urban Plann Dev 129:195–210
Special report: logistics—a moving story. The Economist, December 7, 2002, 65–66
Stiglitz JE (2002) Globalization and its discontents. Penguin Press, London
Thompson A (2005) Factories defuse Chinese challenge by moving up skills and value chains. Financial Times, December 13
Thünen JH (1826) Der isolirte Staat in Beziehung auf Landwirthschaft und Nationalökonomie. Perthes, Hamburg
Timmins N (2010) In the global rush for the new, don't neglect the old. Financial Times, June 8, 2–3
Torrance M (2009) Reconceptualizing urban governance through a new paradigm for urban infrastructure networks. J Econ Geogr 9:805–822
Uchitelle L (2005) Made in the U.S.A. (except for the parts). The New York Times, April 8
Ulanowicz RE, Goerner SJ, Lietaer B, Gomez R (2009) Quantifying sustainability: resilience, efficiency and the return of information theory. Ecol Complex 6:27–36
Vickerman R (2007) Transportation, globalization, and the changing concept of the region. In: Cooper R, Donaghy K, Hewings G (eds) Globalization and regional economic modeling. Springer, Heidelberg, pp 35–43
Wolf M (2004) Why globalization works. Yale University Press, New Haven
Zhang P, Peeta S, Friesz TL (2005) Dynamic game theoretic model of multi-layer infrastructure networks. Netw Spat Econ 5:147–178

Chapter 3
Generating Spatial Time Series on Interstate Commodity Flows

Abstract This chapter presents and demonstrates a methodology for generating spatial time series on interregional (interstate) inter-industry sales (or commodity flows). The methodology embodies an approach to benchmarking and estimating a dynamic multiregional econometric input–output model (or REIM) with annual data, backing out annual interregional inter-industry sales coefficients from the estimated model, and using the coefficients to generate annual observations on commodity flows. The application of this methodology is demonstrated with a REIM that has been estimated for 13 Midwestern, New England, and North Atlantic states and the rest of the United States and 13 industries using time-series data published by the US Bureau of Economic Analysis and the US Bureau of Labor Statistics. The chapter concludes with brief observations on patterns of change in interregional inter-industry commodity flows and sales coefficients.

3.1 Introduction

To model, or test theories about, the evolution of commodity flows and freight movement in the United States in recent decades—as well as its dual phenomenon, the evolution of the geography of production—one needs to generate spatial time series on interregional (or interstate) inter-industry commodity flows to calibrate (econometrically estimate) structural-equation behavioral models. This chapter presents a methodology for generating spatial time series on interstate inter-industry sales, from which inter-industry commodity flows and freight movements may be derived. This methodology builds on the approach taken by Jackson et al. (2006) in constructing commodity-by-industry flow matrices in order to benchmark a dynamic multiregional econometric input–output model (or REIM). In the research here reported we estimate the benchmarked REIM for 13 Midwestern, New England,

Supplementary Information The online version of this chapter (https://doi.org/10.1007/978-3-030-78555-0_3) contains supplementary material, which is available to authorized users.

K. Donaghy et al., *The Co-evolution of Commodity Flows, Economic Geography, and Emissions*, Advances in Spatial Science,
https://doi.org/10.1007/978-3-030-78555-0_3

and North Atlantic states and the rest of the United States and 13 industries using time-series data published by the US Bureau of Economic Analysis and the US Bureau of Labor Statistics. We then back out of the three-region REIM annual interregional inter-industry sales coefficients from which data on interstate inter-industry sales can be generated and freight flows derived and generate the commodity flow data. In the remaining sections of this chapter, we discuss in greater detail our approach to overcoming the lack of spatial time series (in Sect. 3.2), the data used to estimate the three-region REIM and their structure (in Sect. 3.3), the equations of the REIM and their estimation (in Sect. 3.4), an assessment of the estimated REIM (in Sect. 3.5), recovery of interregional inter-industry sales coefficients and trade flows from the estimated REIM (in Sect. 3.6), time profiles of coefficients and flows (in Sect. 3.7), and offer concluding remarks (in Sect. 3.8).

3.2 Overcoming the Lack of Spatial Time Series

To estimate a structural-equation model characterizing the evolution of commodity flows, one would like to have annual observations on, or projections of, interstate commodity flows or inter-industry sales. The Bureau of Transportation Statistics (BTS) publishes estimates of state-to-state commodity flows, obtained from their Commodity Flow Survey (CFS), over multiple year periods. These estimates are not as useful as one would wish for several reasons. One is the fact that all the data of CFS are collected from the origin and destination of shipment instead of the manufacturing location. The desired data, collected on the basis of manufacturing location, would correspond to interstate (interregional) inter-industry sales coefficients updated on an annual basis. Unfortunately, such sales coefficients, when available, are not updated annually. It is for this reason, among others, that the REIM approach to modeling time-varying interregional inter-industry sales was developed. (See Conway (1990) and Israilevich et al. (1997).)

The REIM approach does require matrices of coefficients on interregional inter-industry sales and sales to final demand for a benchmark year, however. Such coefficients are usually obtained from a national or multiregional social accounting matrix (or SAM), whose construction is infrequently attempted because of the difficulty in estimating export distribution (Jackson et al. 2006). The REIM model in this study will employ data extracted from a SAM model developed by Jackson et al. (2006), to be discussed in the next subsection.

3.2.1 *Jackson's et al.'s (2006) Approach to Constructing Commodity-by-Industry Flow Matrices*

In this subsection, we briefly review an alternative methodology for generalizing the distance–volume relationships in the CFS data. Working with national data, Jackson et al. (2006) estimate interregional commodity flow distributions and modify the trade portion of each regional SAM to retain consistency. On their approach, domestic imports must equal domestic exports for all regions.

To achieve the desired consistency, they first estimate an equation to generate the distribution of known regional domestic exports from one region to others in the model. In this estimation, it is assumed that distributions of exports are fixed while export levels can change with regional output levels. As a result, distributions of exports are only a function of transportation cost, measured by interregional distances and regional specific commodity demands, given elasticities of demand with respect to distance and commodity demand levels. Commodities with larger distance elasticities of demand will be more sensitive to variations in demand level while those with smaller demand level elasticities of demand will be more sensitive to shipment distance. The estimation of elasticities should be conducted so as to minimize the absolute difference between estimated and observed commodity flows, generated from the BTS commodity-specific survey data.

3.2.2 *Recovering* I–O *Coefficient Matrices*

In this subsection, we follow Miller and Blair (1985) in discussing computation of interregional inter-industry sales coefficient matrices, to be derived from the SAM produced by Jackson et al. (2006).

Identifying Use and Make Matrices One of the most widely used conventional frameworks for constructing a SAM for a single region is shown in Table 3.1. In this table domestic final demand includes consumption, investment, and government spending, the sum of which is denoted by F, and exports (X) net of imports (M). The Matrices U, V, W, and E are, respectively, the Use, Make, Value Added, and Final Demand matrices, while Q and X correspond to commodity production and total industrial output. The Use matrix represents the industry purchase of each row commodity and the Make matrix represents the commodity output of each row

Table 3.1 Conventional commodity-by-industry framework for a single region

	Commodity	Industry	Final demand	Total
Commodity		U	$E = F + X(M)$	Q
Industry	V			X
Value added		W		
Total	Q'	X'		

industry. Value added includes all payments sectors, whereas final demand represents row commodity final demand by column final demand activity.

Obtaining Total Output by Commodity and Industry from the Use and Make Matrices From Table 3.1, it should be obvious that we can compute elements of the column vector Q by adding the row sums of U and E and compute the elements of the column vector X by taking the row sums of V. To be specific, we can state the corresponding industry output identity as:

$$X_i = V_{i1} + V_{i2} + \ldots + V_{im}, \tag{3.1}$$

$$Q_i = U_{i1} + U_{i2} + \ldots + U_{im} + E_i. \tag{3.2}$$

Recall that the definition of a technical coefficient in the general input–output model is:

$$A_{ij} = Z_{ij}/X_j, \tag{3.3}$$

i.e., A_{ij} defines the total value of output from industry i flowing to industry j for each additional dollar value of industry j's output. The matrix form of this relationship is

$$A = Z(X)^{-1}, \tag{3.3'}$$

where $X^\wedge$ stands for the *diagonalized vector* of total output. Similarly, we can write the following relationship in commodity-by-industry terms:

$$B = U(X)^{-1}. \tag{3.4}$$

Given Eq. (3.4), Eq. (3.2) can be rewritten as:

$$Q = BX + E. \tag{3.2'}$$

Commodity-Based Versus Industry-Based Technology Upon obtaining the total outputs of commodities and industries, one has two options to choose between in obtaining the inter-industry sales coefficient matrix—the commodity-based approach and the industry-based approach. According to the former approach, one assumes that an industry's total output is made up of commodities in fixed proportions and thus one can obtain the industry output proportion as:

$$C = V'(X)^{-1}. \tag{3.5}$$

According to the latter approach, however, one assumes that the total output of a commodity is provided by an industry with a fixed proportion technology and one can obtain the commodity output proportion as:

$$D = V\,(Q)^{-1}. \tag{3.6}$$

Direct Definition of Technology Assumptions Versus By-Product Technology Assumption Under the direct definition of technology assumptions, one can define the commodity-by-commodity direct requirements matrices under the commodity-based technology assumption, *Ac*, as:

$$Ac = BC^{-1}. \tag{3.7}$$

According to Eqs. (3.4) and (3.5), one could also rewrite Eq. (3.7) as:

$$Ac = U(V')^{-1}. \tag{3.7'}$$

Hence, one can define the commodity-by-commodity direct requirements matrix under the commodity-based technology assumption directly in terms of the original make and use matrices. Similarly, one could define the commodity-by-commodity direct requirements matrices under industry-based technology assumption, *Ai*, as:

$$Ai = BD. \tag{3.8}$$

According to the by-product technology assumption, one defines all secondary products as by-products and excludes them from the calculation of the commodity-by-commodity direct requirements matrix. Hence, on this assumption, one can write the commodity-by-commodity direct requirement matrix, *Ab*, as:

$$A_{\mathrm{B}} = [U(V*')](V)^{-1}. \tag{3.9}$$

where V^* is a matrix with only off-diagonal elements of *V*.

Thus far, we have reviewed three different formulae for computing an interregional inter-industry sales coefficient matrix, each corresponding to a different technological assumption—*Ac*, *Ai*, *Ab*.

3.3 Data and Data Structure

In this section, we discuss sectoral data used to estimate our three-region REIM and components of final demand, accounting for the types, amounts, and values of end-use goods and services. Final demand information is usually organized along the lines of the national income identity, according to which aggregate output equals the sum of consumption, investment, government spending, and net exports. The data we use on output and income are classified by region and industry and we will discuss their sources and the computations necessary to obtain desired coefficient matrices.

3.3.1 Final Demand

Gross Regional Product Gross Regional Product (GRP), similar in concept to GDP, accounts for the total amount of value added in the economy for a specific period of time. Generally, data on GRP are available at the regional (state) level from the Bureau of Economic Analysis.

Consumption Since regional consumption data by commodity are generally unavailable, we need to generate regional consumption data on the basis of structural relationships estimated for the national level (assuming that consumption patterns hold at the regional level) and regional data that are available. Regional variables include population levels, income levels, unemployment levels, and the consumer price index. Specific models and results are shown as below. Difficulty in estimating a consumption function for every commodity usually leads one to break consumption down into four categories: autos and auto parts, non-auto durable goods, nondurable goods, and services. (See, e.g., Israilevich et al. (1997).)

Investment Investments are generally classified into three categories: residential, nonresidential, and equipment. For residential and nonresidential investment data, several sources are available. We obtained national data from *Current Construction Reports*, where data regarding national residential and nonresidential put-in-place totals could be found. To allocate investment at the regional level, we needed to prorate the national data by construction permit values, which were be obtained from the *Dodge Construction Potentials Bulletin* and *Manufacturing, Mining, and Construction Statistics* from US Census Bureau. However, before applying the permit values directly, a time series for 2-year moving average had to be computed from residential and nonresidential data. We based proration of investment in equipment on the basis of employment.

Government Spending Government spending is usually divided into two categories: Federal government expenditures made within the region, which can be further divided into civilian and military expenditures, and State and Local government expenditures within the region. Data on all these categories are available from the Bureau of Economic Analysis (BEA).

Net Exports There is no direct source for net export data. Instead, we computed net exports by manipulating the income identity, i.e., $X = Y\text{-}C\text{-}I\text{-}G$, in which X is net exports.

3.3.2 Sectoral Variables

Income Sectoral Income, as one component of personal income, is the total of all the wages and salaries earned by people in a specific industry and is computed according to the same sectoral aggregation scheme as is defined in the input–output table around which the REIM is constructed. Usually, income series can be obtained from the Bureau of Economic Analysis.

In addition to the sectoral income series, one also needs to take into account other income series, most of which can be obtained straightforwardly. One of them is property income, as the income received from personal assets. Others are transfer payments, total payments received from government, and contributions to social insurance. Lastly, we should consider the residential adjustment to income, which is an adjustment that accounts for those people who work and earn wages within the region but do not reside there (Table 3.2).

Table 3.2 Estimates of national-level models of final demand components

Autos and auto parts	Dependent variable	log(Consumption1/Population)
	Coefficient	*t*-Value
Constant	−12.73	−9.72
Log(Income/Population)	0.1	0.36
Log(CPI-auto/CPI)	−9.06	−3.24
Log(Unemployment)	−1.03	−2.06
R^2	0.77	
Non-auto durable goods	Dependent variable	log(Consumption2/Population)
	Coefficient	*t*-Value
Constant	−19.3	−12.8
Log(Income/Population)	2.19	9
Log(CPI-house/CPI)	−19.56	−3.92
Log(Unemployment)	−0.3	−0.64
R^2	0.89	
Nondurable goods	Dependent variable	log(Consumption3/Population)
	Coefficient	*t*-Value
Constant	−15.96	−12.81
Log(Income/Population)	1.41	4.17
Log(CPI-apparel/CPI)	0.94	1.33
Log(Unemployment)	−0.62	−2.07
R^2	0.86	
Services	Dependent variable	log(Consumption4/Population)
	Coefficient	*t*-Value
Constant	−11.08	−7.75
Log(Income/Population)	−0.23	−0.5
Log(CPI-Services/CPI)	4.33	3.49
Log(Unemployment)	−0.89	1.82
R^2	0.85	

All three categories mentioned above, together with the sum of sectoral income, constitute personal income:

$$\begin{aligned}\text{Personal Income} = {} & \text{Total Income} + \text{Property Income} \\ & + \text{Transfer Payments} \\ & - \text{Contributions to Social Insurance} \\ & + \text{Residential Adjustment to Income.}\end{aligned}$$

Employment Employment and unemployment data are available from the Bureau of Labor Statistics (BLS), defined as the total of full-time and part-time jobs.

Output Output data are also generally available from the Bureau of Economic Analysis (BEA). Output data should represent the total values of sales in order to match the input–output table of the base year. For manufacturing industries, output can also be estimated from time-series data on value added, cost of materials, and payroll from Census of Manufacturers and the Annual Survey of Manufacturers. For nonmanufacturing industries, proration should be made with national output series from the Bureau of Labor Statistics according to the employment information.

It should be noted that the Standard Industrial Classification (SIC) System, according to which sectoral data were classified before 1997, has been changed to the North American Industry Classification System (NAICS). Some references are available from US Census Bureau to bridge the two systems and we have followed them in constructing the time series used in estimating the three-region REIM.

3.4 The Three-Region REIM

3.4.1 Brief Description of the Model

The three-region REIM has been estimated with annual data on the economies of 13 Mid-West, New England, and North Atlantic states from 1977 to 2007, organized according to 13 industrial sectors. As a REIM, the model consists of two parts: an input–output part and an econometric part (Tables 3.3 and 3.4).

3.4.2 Expected and Actual Output

In this subsection, we discuss the difference between the output estimations from the embedded input–output model and the REIM, based on Donaghy et al. (2007). Within the three-region model, there are two types of output: actual output, X, and expected output, Z, given by the following relationship:

Table 3.3 Characteristics of the three-region model

Project horizon	1977–2007 (31 years)
Model size	169 sets of linear equation system
Industry detail	13 industries with projections of output and inter-industry/regional commodity flow
Selected exogenous variables	Gross regional products
	Personal consumptions
	Residential and non-residential investment
	Federal government expenditures
	State and local government expenditures
	Population
	Unemployment rate
	Personal income
	Consumer Price Index (CPI)

Table 3.4 Three-region input–output table sector definitions

	Description	NAICS code
1	Agriculture, forestry, fishing, and hunting	11
2	Mining	21
3	Construction	23
4	Food product manufacturing	311
5	Chemical manufacturing	325
6	Primary metal manufacturing	331
7	Fabricated metal product manufacturing	332
8	Machinery manufacturing	333
9	Computer and electronic product manufacturing	334, 335
10	Transportation equipment	336
11	Other nondurable manufacturing	312–316, 322–324, 326
12	Other durable manufacturing	321, 327, 337, 339
13	TCU, services and government enterprises	42, 44, 45, 48, 49, 51–56, 61, 62, 71, 72, 81

$$Z = AX + Y, \tag{3.10}$$

where A is the interregional inter-industry sales coefficient matrix and Y is the vector of final demand. Let us denote the difference between the actual output and expected output by $\Delta = Z - X$. Hence, we could modify Eq. (4.1) as:

$$Z = \Delta + X = AX + Y. \tag{3.11}$$

On the other hand, if the usual input–output approach is used, then:

$$Z_{\text{IM}} = (IA)^{-1}Y. \tag{3.11'}$$

From Eq. (3.11′), it is obvious that:

$$Z_{\text{IM}} = (IA)^{-1}Y = (IA)^{-1}\Delta + X. \tag{3.12}$$

As a result, we could determine the difference between the level of output forecasted by the input–output model and the REIM estimation of output as:

$$Z_{\text{IM}}Z = \left[(IA)^{-1}\,\Delta + X\right][\Delta + X] = \left[(IA)^{-1}I\right]\Delta. \tag{3.13}$$

Using the power series decomposition of the inverse, it obvious that:

$$Z_{\text{IM}} - Z = \left(A + A^2 + \ldots\right)\Delta, \tag{3.14}$$

which clearly shows the difference between estimates made from a traditional input–output model and a REIM, which depends on the structure of A, the structure of the economy as indicated by the nature of the linkages between industries.

In this part of the model, the prediction equations detail the allocation of all commodities produced in 13 states, either consumed as an intermediate good or as a component of final demand, which is divided into consumption, investment, government spending, and net export. Since this process only consists of static equations, the predicted values are usually not equivalent to the actual output except for the base year, which is 2001 in this chapter. We discuss the stochastic adjustment of the estimated output in the next subsection.

3.4.3 Stochastic Equation of Output Adjustment

Having forecasted output on the basis of Eq. (3.10), one needs to account for the stochastic relationship between actual output, expected output, and a set of exogenous variables to turn the model into an econometric forecasting model. In a REIM, this is accomplished through a set of regression equations:

$$\log\left(X_{i,t}/Z_{i,t}\right) = \alpha_0 + \alpha_z(Z_{i,t1}/X_{i,t1}) + \alpha_g G_{it} + \xi_{it} \tag{3.15}$$

for all $i = 1 \ldots n$ and for all $t = 1 \ldots T$, where Z_i, t - 1 is a lagged input–output generated predicted output, and G_{it} is a set of exogenous variable selected by the modeler.[1] Equation (3.15) is estimated separately for each industry i.

[1]Generally, those variables include: Investment, Government Expenditure, and Net Exports.

Assuming that the final demand matrix is exogenously determined, one can write the system of equations implied by (3.15) as follows:

$$\beta_{it} \equiv X_{i,t}/Z_{i,t} = \exp\left(\alpha_0 + \alpha_z(Z_{i,t1}/X_{i,t1}) + \alpha_g G_{it}\right). \tag{3.16}$$

for all $i = 1 \ldots n$ and for all $t = 1 \ldots T$. From Eq. (3.15), it is obvious that:

$$X_i = \beta_i^{Z_i}. \tag{3.17}$$

One can also rewrite the system implied by (3.15) in the form of the following matrix equation:

$$X = \beta^{AX} + \beta^{Y}. \tag{3.18}$$

From which once can obtain the reduced form of the system as:

$$X = \left[\left(\mathrm{I} - \beta^{A}\right)^{-1}\beta\right]Y. \tag{3.19}$$

From the preceding it should be apparent that β, as a function of A, now serves as a multiplier to update the inter-industry sales. As a result, A and β are interdependent. On one hand, as a function of A, β changes as shown in Eq. (4.7). On the other hand, β could also affect A through Eqs. (4.9) and (4.10). As a result, the differences in A on an annual basis can be accounted for by the estimated β.

To this point, we have discussed the approaches to obtain expected output and the dynamics of output adjustment. We next turn to the specification and estimation of the model's other equations.

3.4.4 Stochastic Equations of Productivity

As in the case of output, the employment equation characterizes the relationship between an industry's total shipments and total employment, which is defined as:

$$\log\left(X_{i,t}/N_{i,t}\right) = \alpha_0 + \alpha_z(N_{i,t1}/X_{i,t1}) + \alpha_g G_{it} + \xi_{it} \tag{3.20}$$

for all $i = 1 \ldots n$ and for all $t = 1 \ldots T$, where *Zi*, *t* - 1 is a lagged input–output generated predicted output, and *Git* is a set of exogenous variable selected by the modeler, while *Nit* is the employment of sector *i* at the time *t*.

3.4.5 Stochastic Wage Equation

The final equation combines the total earnings with the total employment to obtain the estimations of personal income.

$$\log\left(Y_{i,t}/N_{i,t}\right) = \alpha_0 + \alpha_z(N_{i,t1}/Y_{i,t1}) + \alpha_g G_{it} + \xi_{it} \tag{3.21}$$

for all $i = 1 \ldots n$ and for all $t = 1 \ldots T$, where *Zi*, *t* - 1 is a lagged input–output generated predicted output, and *Git* is a set of exogenous variable selected by the modeler, while *Nit* is the employment of sector *I* at the time *t* with *Yit* as the income of sector *i* at time *t*.

3.4.6 Model Estimation and Solution

For each industry, the four equations can be estimated separately with different structural econometric specifications for each, though usually a systems or nonlinear systems estimator is employed in estimating parameters of REIMs (Donaghy et al. 2007). In the present case, the equations of the model were econometrically estimated with a nonlinear generalized least-squares estimator (NLGLS) implemented in the statistical package R. The four equations are closely interrelated. To be specific, predicted output and actual output are interdependent, which could be verified from the first two equations. Moreover, employment is dependent on output and income depends on employment. Hence, the systems of equations for all industries are solved simultaneously to obtain the forecasted values of the endogenous variables.

3.5 Model Assessment

Generally speaking, the model—all sets of equations estimated—performed (fit the data) quite well. In this section, we offer an assessment of the estimated REIM through a series of statistical diagnostics for Eqs. (3.15), (3.20), and (3.21) selected for a few industries and states.

In Chart 3.1, there are four plots of the values of variables and residuals corresponding to the dynamic output adjustment Eq. (3.15) in the case of the "other durable manufacturing" sector of Connecticut. For the fitted values, it is quite obvious from the plot in the upper-left quadrant that all the estimated values from 1976 to 2007 are between −0.2 and 0.2. And, since the predicted values are in the logarithm of the ratio of estimated output and actual output, the results indicate that the error ratios of estimations, obtained from (3.15), are within 20%. Displayed in the upper-right plot is the relationship between standardized residuals and fitted

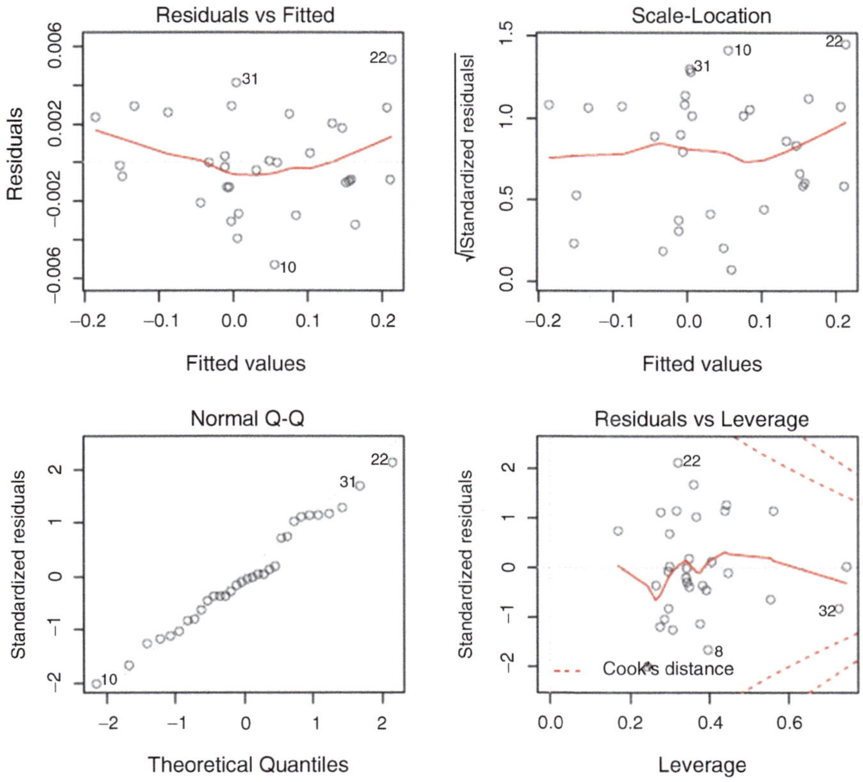

Chart 3.1 Regression diagnostics for output of other durable manufacturing in Connecticut

values, from which no distinctive relationship is detected. From this plot, we may reasonably infer that the model has taken into account almost every relevant explanatory factor for the adjustment of output of this industry over the sample period.

The plot in the lower-left quadrant, the QQ-plot, is usually used to compare two distributions of residuals, where complete randomized standard-normal distribution data are represented by a dashed straight line. The linearity of the points for residuals suggests that the residuals are normally distributed. The plot in the lower-right shows what is termed Cook's distance, which is a measure introduced by Cook (1977) that is commonly used to estimate the influence of a data point. Since all the distances shown in the plot are smaller than 0.5, it can be inferred that there are no influential outliers in the data.

From Chart 3.2 which displays regression diagnostics for the productivity Eq. (3.20) of the "TCU, services, and government enterprises" sector in Wisconsin, it is also quite obvious that no specific trend is detected between fitted values and residuals. Furthermore, the QQ plot also indicates that the residuals are in

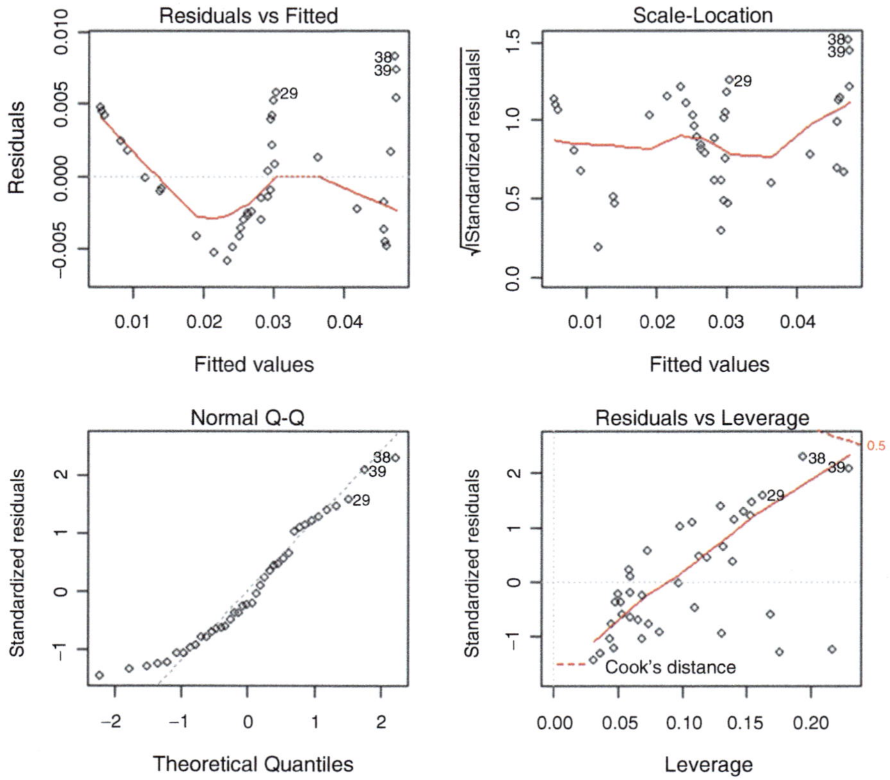

Chart 3.2 Regression diagnostics for productivity of TCU, services, and government enterprises in Wisconsin

conformance with the normal distribution, while all the standardized residuals' Cook's distances are smaller than 0.5, indicating no influential outliers.

One can draw similar conclusions from Chart 3.3 which displays regression diagnostics for the wage Eq. (3.21) of the "TCU, services, and government enterprises" sector in Wisconsin.

3.6 Recovering Annual Interregional Inter-industry Sales Coefficients from the Estimated REIM

3.6.1 Derivation of I–O *Coefficient Matrix and Leontief Inverse*

In the foregoing, we have discussed the specification and estimation of a REIM system in which the relationship between estimated and observed outputs has been

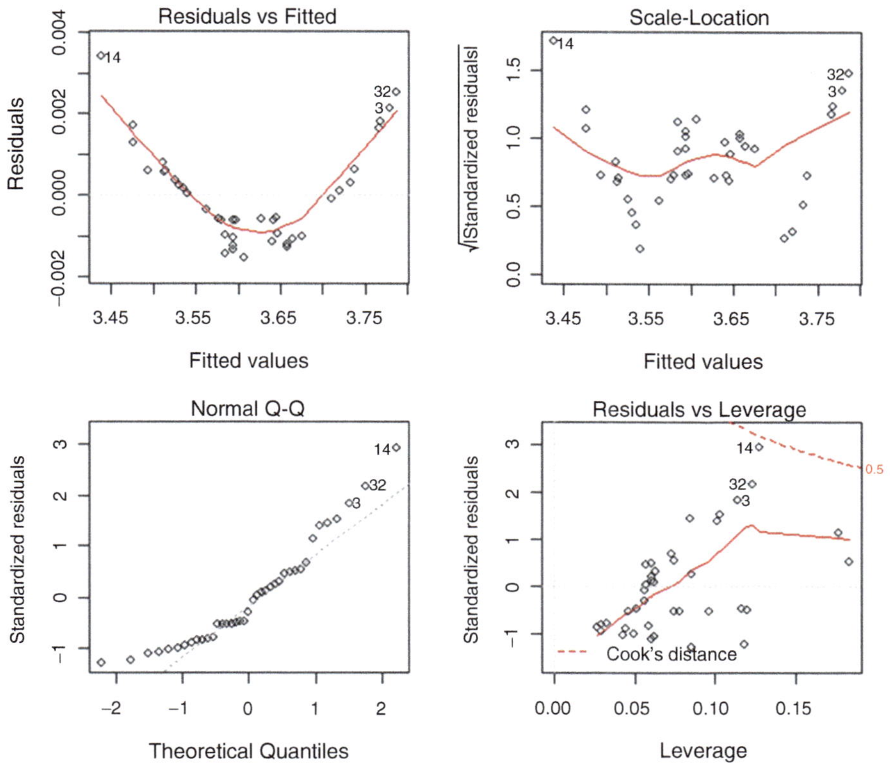

Chart 3.3 Regression diagnostics for the wage equation of TCU, services, and government enterprises in Wisconsin

established. In a traditional *I*–*O* approach, the inter-industry sales matrix, *A*, will generally be estimated from the inter-regional trade matrix initially and then the Leontief inverse, *B*, will be computed as $(I–A)^{-1}$. In the present situation, in which we seek an approximation of *A*, however, the estimation process will be reversed with the Leontief inverse being estimated as the first step and then the interregional inter-industry sales matrix derived secondly. In the Leontief inverse, each element, e_{ij}, is an output multiplier indicating the amount by which industry *i* will increase its sales to industry *j* and every other sector in the economy when sector *j* delivers one more unit of output to final demand. While computation of the elements of the Leontief inverse for non-benchmark years is complicated, it can be accomplished by following the steps provided by Israilevich et al. (1997).

Equations (3.20) and (3.21) offer us a direct relationship between income, output, and final demand. With that relationship in mind, it is straightforward to determine the relationship between income, or final demand and output. Then, the required relationship can be fed back to the *I*–*O* Eq. (4.7) to compute the inter-industry gross output correlation, which can be regarded as the Leontief inverse. The interregional inter-industry sales coefficient matrix can then be easily obtained by reversing the steps set out in the fourth section. We must observe that this approach will yield at best an approximation of A and corresponding trade flows. Given the quality of the data on which this work is based, it is unlikely that the imputed interregional inter-industry sales coefficient matrix and Leontief inverse matrix will ever be identical to the true but unknown matrices and, unfortunately, there are no effective approaches to assess the accuracy of the computations described above (Israilevich et al. 1997).

From Eqs. (3.20) and (3.21), we can derive the following relationships:

$$\gamma_{it} \equiv X_{i,t}/N_{i,t} = \exp\left(\alpha_0 + \alpha_z(N_{i,t1}/X_{i,t1}) + \alpha_g G_{it}\right), \tag{3.22}$$

$$\mu_{it} \equiv Y_{i,t}/N_{i,t} = \exp\left(\alpha_0 + \alpha_z(N_{i,t1}/Y_{i,t1}) + \alpha_g G_{it}\right), \tag{3.23}$$

which imply the following two relationships:

$$N = (\gamma)^{-1}X. \tag{3.24}$$

$$Y = (\mu)N. \tag{3.25}$$

Substitution of (3.24) into (3.25) gives the following relationship between final demand and total output:

$$Y = (\mu)\,(\gamma)^{-1}X. \tag{3.26}$$

With (3.26) in hand, the Leontief inverse matrix of output multipliers can be written in the following form:

$$B = \left[(I\beta^A)^{-1}\beta\right](\mu)(\gamma). \tag{3.27}$$

Given annual estimates of the Leontief inverse matrix B_t, annual estimates of the interregional inter-industry sales coefficients matrix A_t can be derived as:

$$A_t = IB_t^{-1}, \tag{3.28}$$

and annual interregional inter-industry trade flows can be computed from annual regional (state) aggregate output levels as:

$$x_{ij_t}^{lm} = a_{ij_t}^{lm} X^m_{j_t}, \forall i, j, l, m, \tag{3.29}$$

where $x_{ij_t}^{lm}$ denotes the sales of industry i in region (state) l to industry j in region (state) m in year t, $X^m_{j_t}$ denotes aggregate output of sector j in region (state) m in year t, and $a_{ij_t}^{lm}$ is the corresponding interregional (interstate) inter-industry sales coefficient for year t. Physical characteristics of the shipments—volume and weight—can then be estimated in the usual manner from industry coefficients (Boyce 2002; Ham et al. 2005).

3.7 Time Profiles of Trade Flows and Sales Coefficients

In this section, we offer some general observations about, and illustrate changes in, estimates of interregional inter-industry trade flows and sales coefficients over the sample period, 1977–2007. (Commodity flows and physical characteristics of the associated shipments derived from this model are analyzed in greater detail in Chap. 4.)

We observe that, while interregional inter-industry sales coefficients increased for many of the industries, indicating a closer connection between states and industries, trade between some industries and states declined noticeably, with the mining industry in Midwestern states being a leading example. As a counterexample, the fabricated metal product manufacturing industry exhibited a strong growth trend in sales to other sectors and regions, except for these sectors in Indiana and Ohio. As one might expect of the sample period, the computer and electronic product manufacturing sector, wherever it was located, also increased sales to other regions and sectors, except in the case of New York.

As noted in the introduction, one of the important developments in trade in the last few decades has been the increase in intra-industry trade. In the next two charts provided below, we plot the volumes of same-sector trade between Illinois and Connecticut. In Charts 3.4 and 3.5, the trade volumes are measured along the vertical axis and time on the horizontal one. In these charts and those ensuing sectoral plots are aligned in four columns corresponding to sectors 1–4, 5–8, 9–12, and 13.

From the plots in these charts it is evident that there is little, if no, trade between firms in some sectors, increasing trade in many, and decreasing trade in others. What is also obvious is that the imputed changes in interregional inter-industry trade are large enough in relative terms to warrant efforts to estimate them and to attempt to explain them.

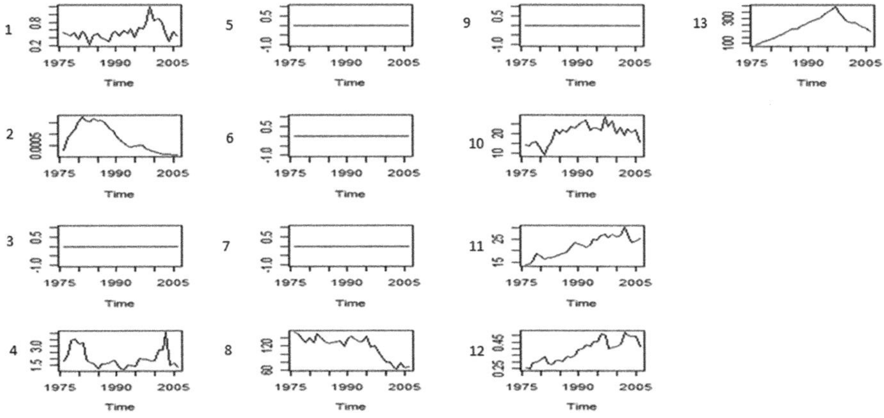

Chart 3.4 Trade volumes for same-sector interstate trade from Connecticut to Illinois for all 13 industries

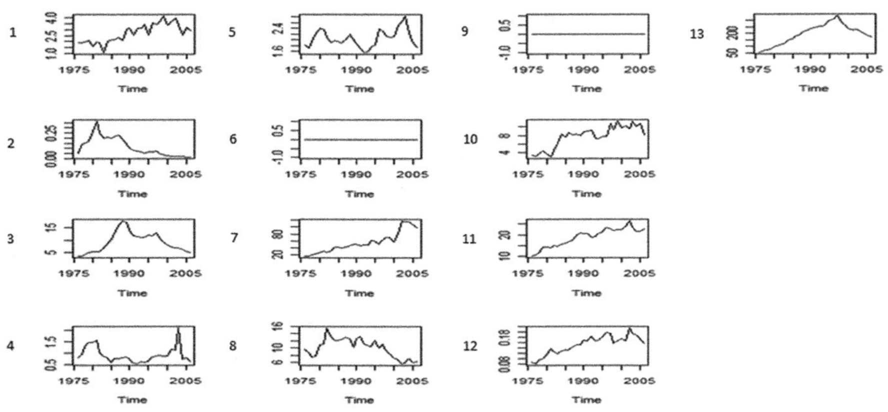

Chart 3.5 Trade volumes for same-sector interstate trade from Illinois to Connecticut for all 13 industries

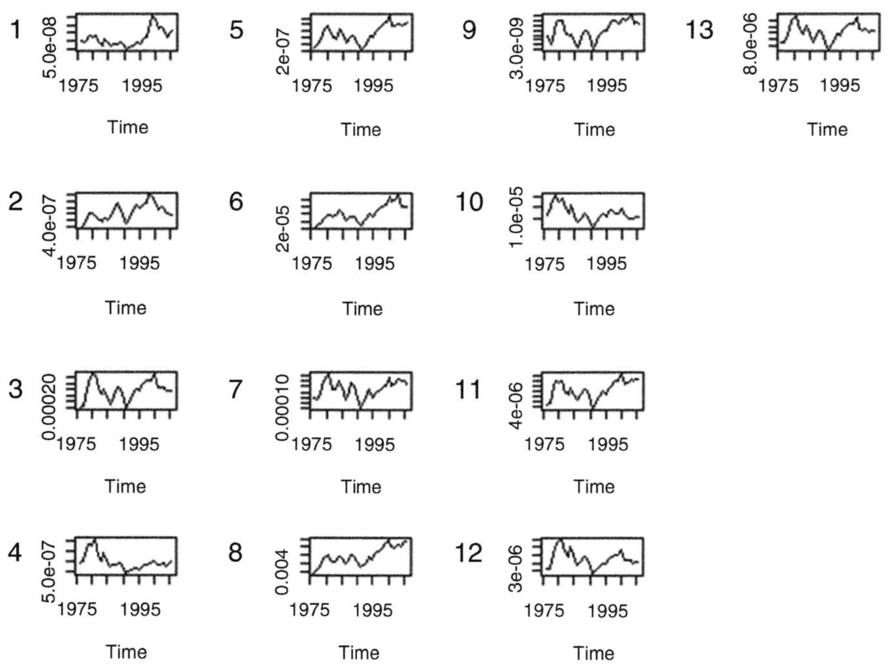

Chart 3.6 Time profile for interregional inter-industry sales coefficients for sales from machinery manufacturing in Illinois to all industries in Indiana

We now turn our attention to changes in the interregional inter-industry sales coefficients and consider how trade from one regional (state) sector to all other sectors in another region has evolved and vice versa. In the following charts the values of interregional intra-industry sales coefficients are measured on the vertical axis (Charts 3.6 and 3.7).

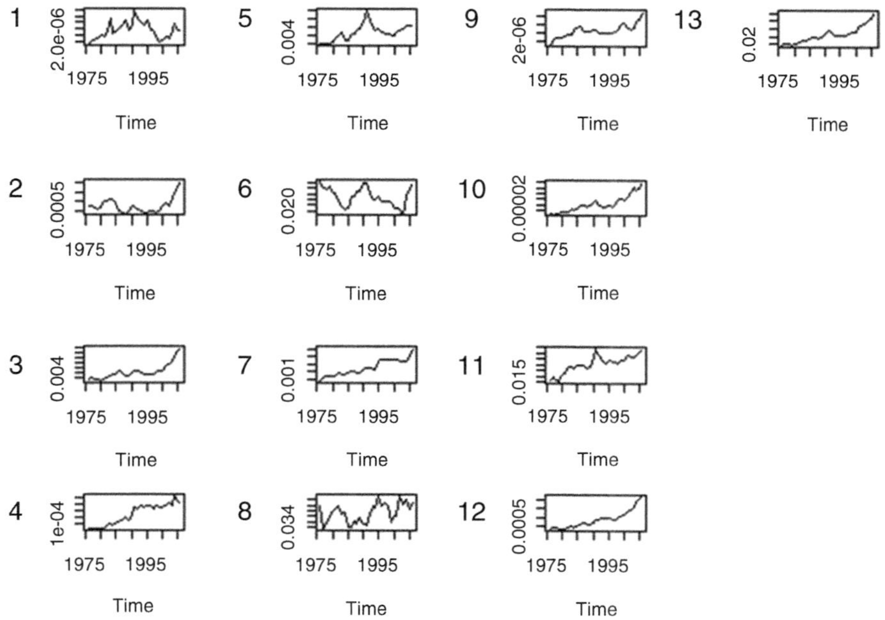

Chart 3.7 Time profiles of interregional inter-industry sales coefficients for sales from all industries in Indiana to machinery manufacturing in Illinois

From these plots of time varying intra-industry sales coefficients it is difficult to draw general conclusions about the nature of evolving interregional inter-industry sales involving the industries considered, even as coefficients for the most part are trending upwards. Indeed, arriving at such conclusions is the objective of the next stage of the analysis in this research project. (Please see Appendix for other time profiles of interregional interindustry sales coefficients.)

3.8 Conclusions

In the foregoing sections of this chapter, we have elaborated and demonstrated a methodology for generating spatial time-series data on interregional inter-industry sales and associated commodity or freight flows. These data will be made available to the wider research community through a Springer website associated with this volume.

These data can be employed in econometrically estimating dynamic structural-equation behavioral models characterizing the evolution of interregional commodity flows (as in Chap. 6). Such models should prove useful in testing theories about the evolution of freight flows and the economic geography that is dual to it. Such models should also prove useful in forecasting what the effects on infrastructure, air quality, and nodal communities increasing trade may have (as in Chap. 7) and conducting thought experiments about what appropriate policy responses may be.

Acknowledgments The authors would like to express their deep gratitude to Randall Jackson, Gianfranco Piras, and Geoffrey Hewings, without whose invaluable assistance this chapter would not have been written.

Appendix: Other Time Profiles of Interregional Inter-Industry Sales Coefficients

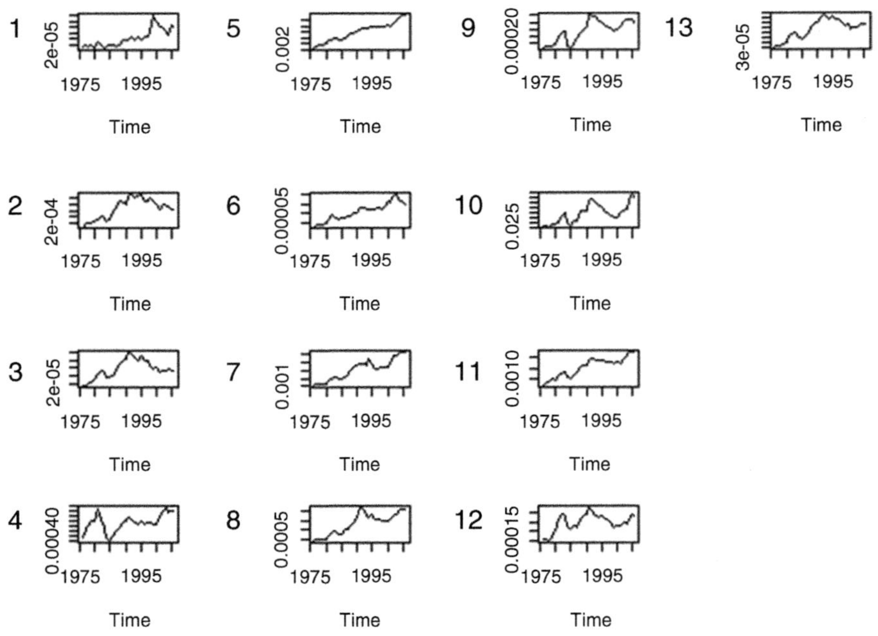

Chart 3.8 Time profiles of interregional inter-industry sales coefficients for sales from chemical manufacturing in Ohio to all industries in Illinois

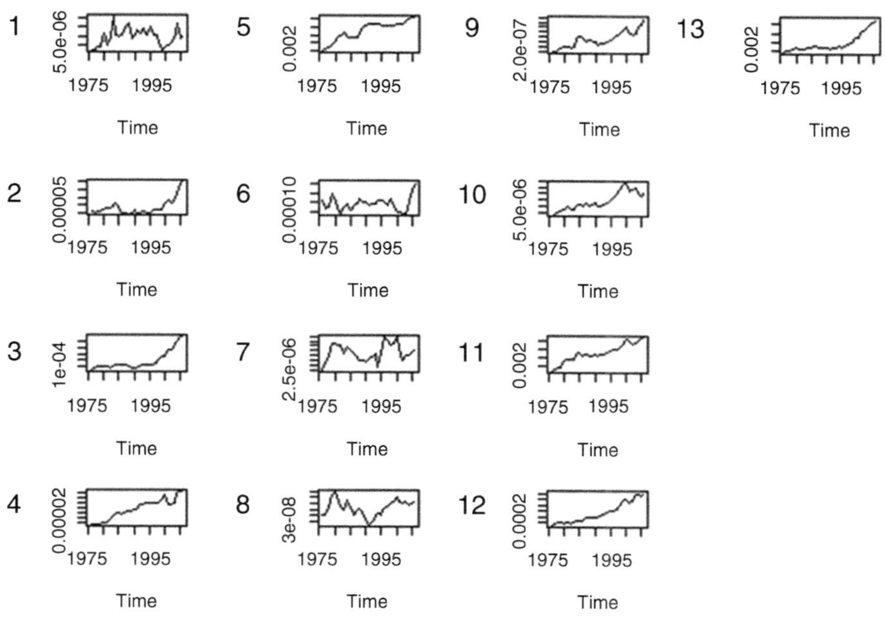

Chart 3.9 Time profiles of interregional inter-industry sales coefficients for sales from all industries in Illinois to chemical manufacturing in Ohio

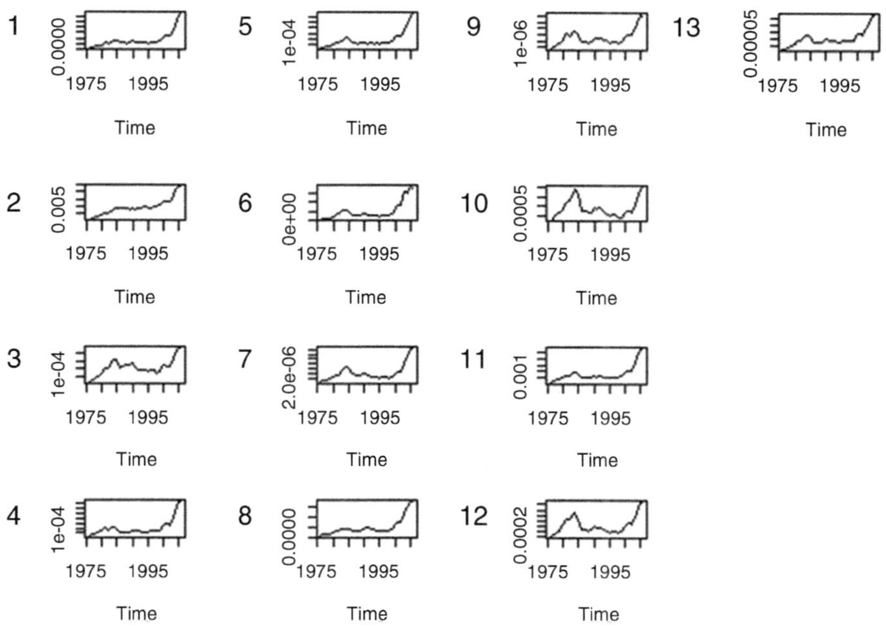

Chart 3.10 Time profiles of interregional inter-industry sales coefficients for sales from mining in Pennsylvania to all industries in New York

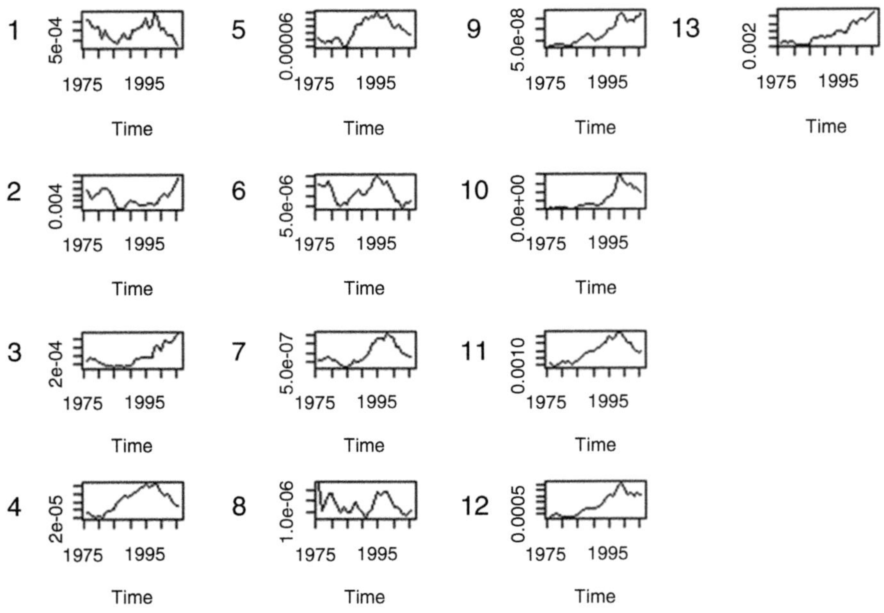

Chart 3.11 Time profiles of interregional inter-industry sales coefficients for sales from all industries in New York to mining in Pennsylvania

References

Boyce D (2002) Combined model of interregional commodity flows on a transportation network. In: Hewings GJD, Sonis M, Boyce D (eds) Trade, networks, and hierarchies: modeling regional and interregional economies. Springer, Heidelberg, pp 29–40

Conway RS (1990) The Washington projection and simulation model: ten years of experience with a regional interindustry econometric model. Int Reg Sci Rev 13:141–165

Cook R (1977) Detection of influential observations in linear regression. Technometrics 19:15–18

Donaghy KP, Balta N, Hewings GJD (2007) Modeling unexpected events in temporally disaggregated econometric input-output models of regional economies. Econ Syst Res 19:125–145

Ham H, Kim TJ, Boyce D (2005) Implementation and estimation of a combined model of interregional multimodal commodity shipments and transportation network flows. Transp Res B 39:65–79

Israilevich P, Hewings GJD, Sonis M, Schindler GR (1997) Forecasting structural change with a regional econometric input-output model. J Reg Sci 37:565–590

Jackson RW, Schwarm WR, Okuyama Y, Islam S (2006) A method for constructing commodity by industry flow matrices. Ann Reg Sci 40:909–920

Miller R, Blair P (1985) Input-output analysis: foundations and extensions. Prentice-Hall, Englewood Cliffs, NJ

Chapter 4
The Coevolution of Commodity Flows, Economic Geography, and Associated NonPoint Source Black Carbon Emissions in the Midwest-Northeast Transportation Corridor of the United States, 1977–2007

Abstract The previous chapter of this volume discussed the generation of spatial time-series data on interstate inter-industry trade flows for the Midwestern and Northeastern states of the United States and the rest of the country over the period of 1977 to 2007. This chapter provides an analysis of these data and detailed commentary on changes in aggregate volumes of shipments (reflecting the increasing transport intensity of production and consumption), changes in patterns of intra-industry shipments (reflecting changes in economic geography), and changes in patterns of associated black carbon emissions that result from the movement of goods.

4.1 Data Analysis

Solution of the regional econometric input–output model, discussed in Chap. 3, for annual interstate inter-industry trade flows over the period of the sample by the approach of Israilevich et al. (1997), yields 2366 time series—13 sectors × 13 sectors × 13 states plus the rest of the United States. We will therefore not attempt to offer here a comprehensive characterization of either the evolution of the commodity or freight flows or a commentary on regions that have come to specialize in activities involving various trade activities. We will, however, comment on aggregate changes in volumes of shipments, intra-industry shipments, and associated black carbon emissions that result from the movement of goods. Physical characteristics of the shipments—volume and weight—can be estimated from sales figures in the usual manner from industry coefficients (Boyce 2002; Ham et al. 2005).[1]

The principal author and lead researcher of this chapter was Benjamin Brown-Steiner. The analysis is based on data derived by Jialie Chen.

---

[1]Conversion factors in millions of constant year-2001 US dollars per kiloton for the 13 sectors are respectively 0.888524, 0.262133, 0.181015, 1.403953, 2.012055, 1.180493, 0.061437, 9.415343, 22.01849, 7.725906, 1.560004, 1.631441, and 3.103150 (USDOT CFS for 2007).

K. Donaghy et al., *The Co-evolution of Commodity Flows, Economic Geography, and Emissions*, Advances in Spatial Science,
https://doi.org/10.1007/978-3-030-78555-0_4

4.1.1 Changes in Volumes of Shipments

Presented in Tables 4.1a, b and 4.2a, b are 5-year averages of volumes of freight shipments (in kilotons) as inputs and outputs by industries within the 13 states under consideration and by these states and the rest of the United States. Emerging from the data are trends indicating that there are cycles around trends and differential changes—both positive and negative—in the amounts of freight being shipped to firms in industries in different states. States shipping more freight over time include Connecticut, Illinois, New Jersey, New York, Wisconsin, and the rest of the United States. States/regions shipping less freight include Maine, Massachusetts, Michigan, Ohio, and Pennsylvania. Clearly, there is a mix of Mid-western and Northeastern/ Mid-Atlantic states in both groups. From Table 4.1a, b, one can see that there is about a fivefold increase in inputs by weight and about a halving of outputs by weight, indicative of production processes that are increasingly weight-losing or substitution of foreign or extra-regionally produced inputs for regional inputs. While agriculture appears to be the only sector in which both input and output shipments by weight increased over the sample period, sectors 8 through 13 (in the aggregated scheme employed) experienced significant increases in input volumes from other states while reducing output volumes, also indicative of the growing importance of supply chains. Even as markers of globalization appear in the data, it is interesting to observe that the rate of growth in freight movement between firms in sectors in the 13 states considered in this study is smaller than the national trend. This observation suggests that much freight movement through the Midwest-Northeast corridor must constitute pass-through shipments.

4.1.2 Intra-Industry Trade

As discussed thoroughly and insightfully in Munroe et al. (2007) and Hewings and Parr (2009), much regional and inter-regional trade is now intra-industrial in nature because of the amount of fragmentation occurring. The importance of intra-industrial trade (IIT) to a sector is often characterized by computing the associated Grubel-Lloyd index (Grubel and Lloyd 1975).[2] (Please see the Appendix.). One would expect to find clustering of activities in sectors in which there is much IIT as economies of scale and scope are exploited by firms, even as we are unable to detect such phenomena at the level of sectoral and spatial aggregation at which we are

[2]If the Grubel-Lloyd index assumes a value of 1, there is a large amount of intra-industry trade—e.g., in our case, a region imports about the same quantity of a sectoral produce from another region as it exports to that region. If the index assumes a value of 0, there is no interregional intra-industry trade—i.e., a region in consideration only either exports or only imports a good produced by a particular sector.

Table 4.1 Interstate input and output shipments by sector (in kilotons)

	1977–1982	1983–1987	1988–1992	1993–1997	1998–2002	2003–2007
	(a) Sector 5-year averages, input					
agricu	156,894.619	211,203	189,299.508	216,616.613	262,408.292	336,512.415
mining	281,401.586	579,733.373	388,537.932	390,155.424	540,563.975	941,356.298
constr	318,702.205	468,246.616	617,106	685,157.445	992,065.1	1,469,286.38
foodpm	62,453.89	84,227.7851	108,276.937	138,679.389	154,302.386	183,309.218
chemmf	117,139.044	161,338.439	202,729.28	263,658.059	337,059.336	409,305.751
pmetal	150,870.383	155,495.221	201,458.958	247,011.548	285,832.205	311,288.595
fmetal	412,121.495	496,212.301	614,526.343	834,323.093	1,117,692.12	1,106,692.62
machin	18,781.0408	24,046.9963	27,348.3745	38,101.77	62,688.14	75,538.104
comput	339.6206	633.3861	863.5891	1515.889	3793.921	3277.22
treqpt	1620.2772	2012.3913	3348.0436	4679.373	7084.076	8341.005
ondrmf	303,272.107	428,800.19	536,271.157	735,668.786	1,016,411.78	1,140,333.38
odrmfr	25,143.492	33,915.2023	44,514.9534	65,122.443	105,494.548	139,021.107
govten	2,304,594.43	3,387,510.06	4,388,639.4	5,036,037.13	4,674,053.47	6,234,255.57
Total	4.15E + 06	6.03E + 06	7.32E + 06	8.66E + 06	9.56E + 06	1.24E + 07
	(b) Sector 5-year averages, output					
agricu	164,660.894	208,487.95	240,578.35	272,523.92	236,766.48	260,166.92
mining	552,091.647	957,604.88	649,053.08	612,929.47	678,288.66	1,281,683.48
constr	1,055,728.62	1,687,541.52	2,251,582.71	2,489,337	2,796,446.08	4,094,909.19
foodpm	193,628.551	268,083.3	356,178.58	426,909.04	402,963.55	479,720.11
chemmf	68,246.369	99,402.39	140,789.19	177,392.28	195,020.75	255,588.58
pmetal	108,934.912	94,978.02	130,738.67	147,626.48	127,824.3	165,033.45
fmetal	1,541,435.72	1,990,602.91	2,558,613.42	3,300,281.45	3,812,386.46	4,161,612.53
machin	16,880.826	18,851.13	22,198.57	27,859.33	30,822.15	31,940.12
comput	5787.784	10,323.1	13,912.97	18,627.95	21,006.22	19,398.97

(continued)

Table 4.1 (continued)

	1977–1982	1983–1987	1988–1992	1993–1997	1998–2002	2003–2007
treqpt	13,177.999	22,146.79	31,924.16	43,683.8	57,349.16	69,606.83
ondrmf	239,424.114	365,293.19	482,734.27	588,346.59	650,075.64	822,580.63
odrmfr	48,434.42	68,487.84	98,905.78	125,649.87	128,189.77	156,267.08
govten	144,902.339	241,571.94	345,710.71	425,559.75	422,310.11	560,009.77
Total	4.15E + 06	6.03E + 06	7.32E + 06	8.66E + 06	9.56E + 06	1.24E + 07

Table 4.2 Freight shipments received and shipped by state and rest of the United States (in kilotons)

	1997–1982	1983–1987	1988–1992	1993–1997	1998–2002	2003–2007
	(a) State 5-year averages, input					
CT	51,086.71	63,233.03	86,089.75	128,793.7	162,496.3	215,596.5
IL	52,704.74	68,328.05	94,231.03	138,167.8	176,577.3	237,779.6
IN	54,573.53	69,877.52	95,695.79	139,385.3	177,856.8	229,036.5
ME	51,360.58	63,871.27	86,314.23	127,359	159,098.7	204,762.2
MA	50,479.46	61,992.48	84,615.57	126,714.6	160,436.6	215,646.6
MI	44,228.41	57,047.69	78,827.06	118,009.3	153,240.7	201,819.5
NJ	49,074.75	61,768.4	84,282.74	126,241.6	162,575.2	224,852.1
NY	41,691.6	53,896.5	76,288.29	116,186.1	154,928.5	219,920.9
OH	48,755.48	63,169.18	87,323.54	130,847.9	167,597.5	220,501
PA	45,523.56	58,853.39	81,759.86	122,600.9	162,200	219,224.3
RI	55,042.04	68,089.53	92,489.24	134,506.5	164,376.4	214,076.4
VT	53,184.17	66,456.44	90,146.82	131,553	162,704.4	210,817.7
WI	48,777.33	62,159.57	85,370.68	124,987	160,132.2	209,189.5
US	3,506,851.84	5,214,631.91	6,199,485.87	6,991,374.4	7,435,228.8	9,535,294.9
Total	4.15E + 06	6.03E + 06	7.32E + 06	8.66E + 06	9.56E + 06	1.24E + 07
	(b) State 5-year averages, output					
CT	26,940.41	34,844.39	47,241.43	72,078.95	101,769.51	148,415.11
IL	82,483.86	126,347.82	167,726.36	241,036.33	362,108.79	542,696.68
IN	101,792.7	44,218.61	194,108.25	277,195.82	397,832.32	522,450.62
ME	16,786.19	22,108.23	29,380.24	42,514.47	58,247.4	84,621.35
MA	24,494.29	32,337.08	45,227.36	69,212.26	101,770.54	158,529.02
MI	232,275.39	303,275.53	407,351.11	597,092.24	814,823.27	1,000,041.76
NJ	44,514.04	60,809.83	81,457.62	126,534.31	195,665.96	305,502.05
NY	37,280.56	55,830.44	82,982.04	129,510.42	207,004.97	333,876.1

(continued)

Table 4.2 (continued)

	1997–1982	1983–1987	1988–1992	1993–1997	1998–2002	2003–2007
OH	107,949.09	152,551.83	209,143.19	318,866.28	455,034.02	582,048.17
PA	42,342.32	66,209.24	92,832.65	136,290.39	207,074.37	322,861.02
RI	16,988.04	21,507.97	29,320.31	41,277.18	54,442.06	79,289.49
VT	17,652.71	22,587.94	29,671.6	41,200.99	54,787.88	75,197.15
WI	58,282.03	78,568.12	104,653.53	146,616.34	207,987.7	292,692.82
US	3,343,552.57	4,912,177.94	5,801,824.79	6,417,300.99	6,340,900.54	7,910,296.33
Total	4.15E + 06	6.03E + 06	7.32E + 06	8.66E + 06	9.56E + 06	1.24E + 07

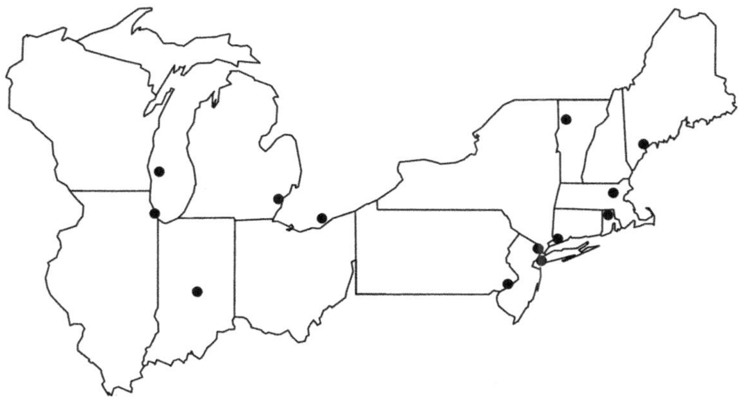

Fig. 4.1 US states in the midwest–northeast transportation corridor

working and by the descriptive approach we are taking. (But see Chen and Donaghy (2012) for an analysis that does find significant scale economies.)

Figure 4.1 provides a map of US states in the Midwest-Northeast transportation corridor (with their largest cities located within them) onto which bidirectional trade flows will be gridded.

In Fig. 4.2a–f, we provide spatial plots of sectoral IIT (in both directions) and the plots over time of the Grubel-Lloyd indices for six of the 13 industries studied. (The values of the Grubel-Lloyd index are measured on the vertical axis.) These plots illustrate a number of salient points about interstate IIT.

- For some sectors and between some states there is considerable variation in IIT.
- IIT is very limited across all states in some industries in which there is specialization—e.g., transportation equipment—but also fairly high across many states in other industries—e.g., other nondurable manufacturing.
- There have not been dramatic changes in IIT patterns over the 31-year period of the sample, certainly less so than we may have expected; but it may be the case that IIT increased significantly in the late 1960s and 1970s before it was closely studied.
- Some regional trade patterns in the Midwest-Northeast corridor have emerged as Munroe et al. (2007) observed more narrowly within the Midwest—in the cases of agriculture, food product manufacturing, and TCU, Services, and Government Enterprises—while corridor-wide patterns have emerged in others—in the case of other nondurable manufacturing—and virtually not at all in others—e.g., in the case of transportation equipment.

From Fig. 4.2a, e, it appears that between 1977 and 2007 interregional intra-industry shipments of agricultural goods and nondurable manufactured goods were highest. From Fig. 4.2d, f it would appear that over this period interregional intra-industry shipments of transportation equipment and products of the TCU, services, and government enterprises sector were the lowest.

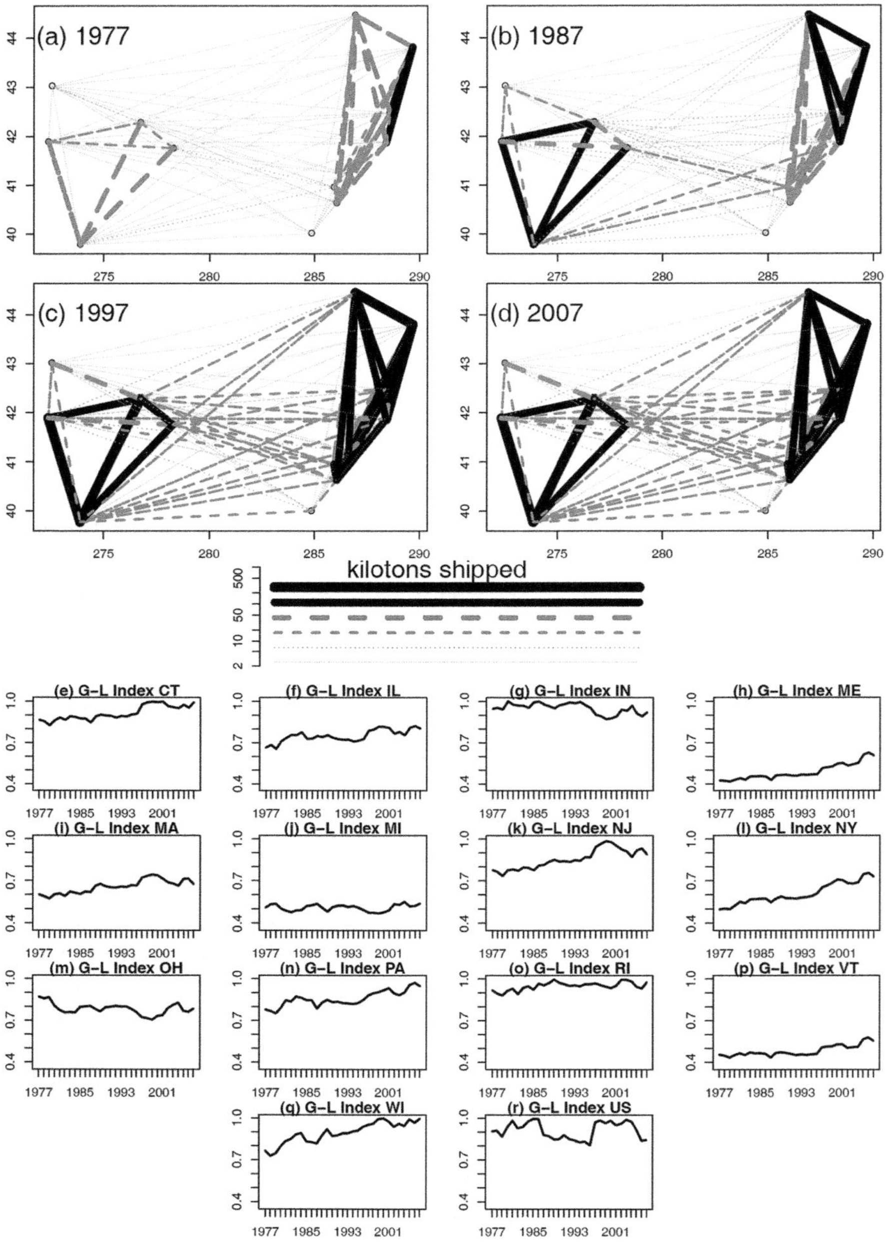

Fig. 4.2 (**a**) Intra-industry shipments for agriculture and state Grubel–Lloyd indices. (**b**) Intra-industry shipments for food product manufacturing and state Grubel–Lloyd indices. (**c**) Intra-industry shipments for fabricated metal product manufacturing and state Grubel–Lloyd indices. (**d**) Intra-industry shipments of transportation equipment and state Grubel–Lloyd indices. (**e**) Intra-industry shipments of other nondurable manufacturing and state Grubel–Lloyd indices. (**f**) Shipments of TCU, services, and government enterprises and state Grubel–Lloyd indices

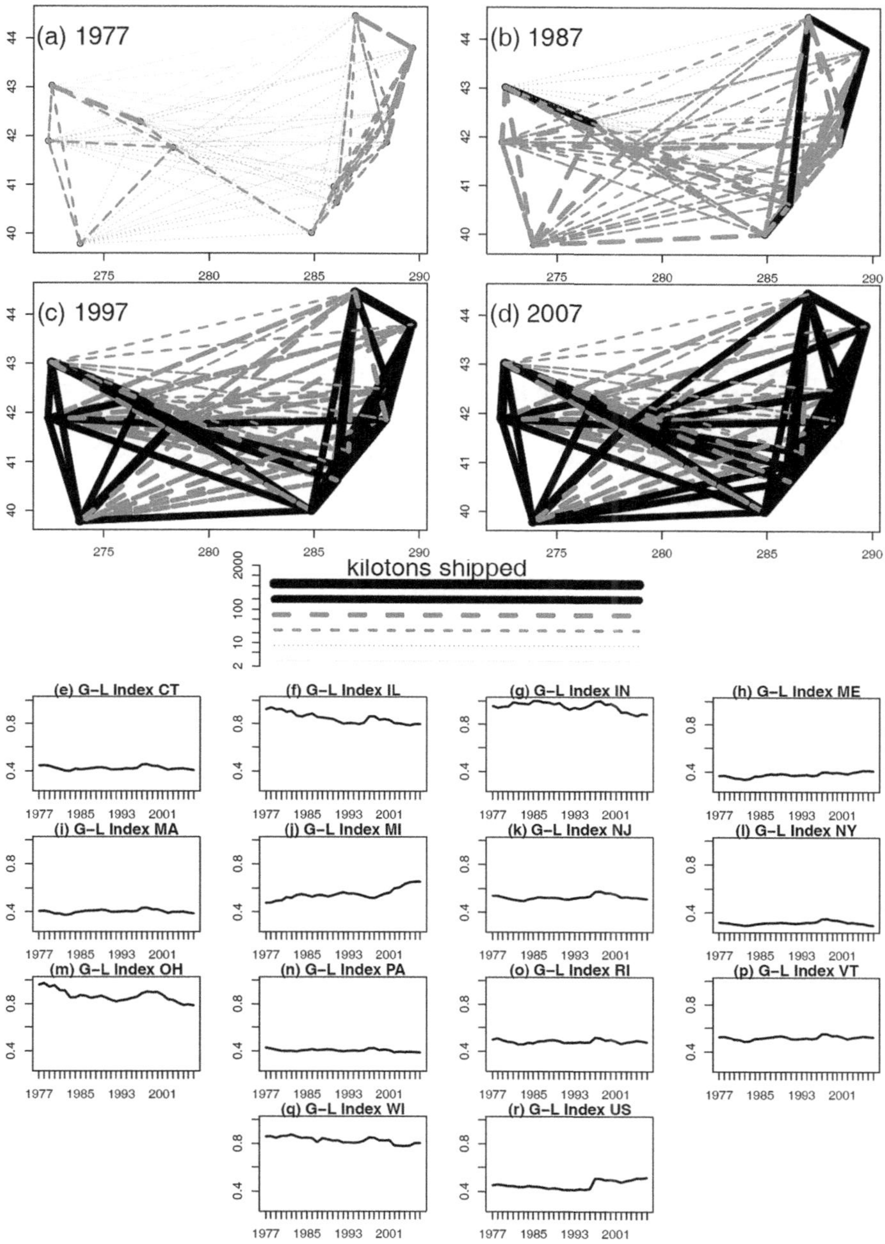

Fig. 4.2 (continued)

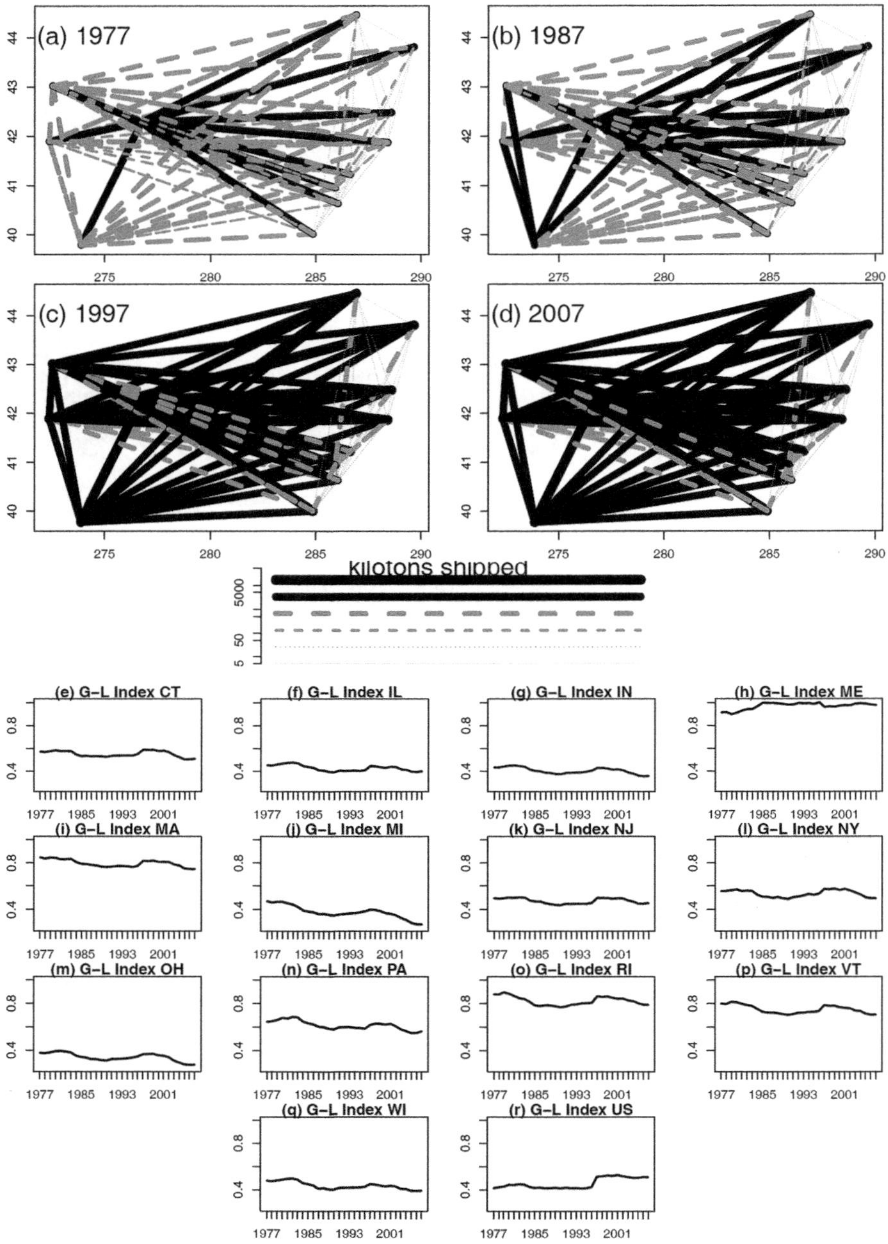

Fig. 4.2 (continued)

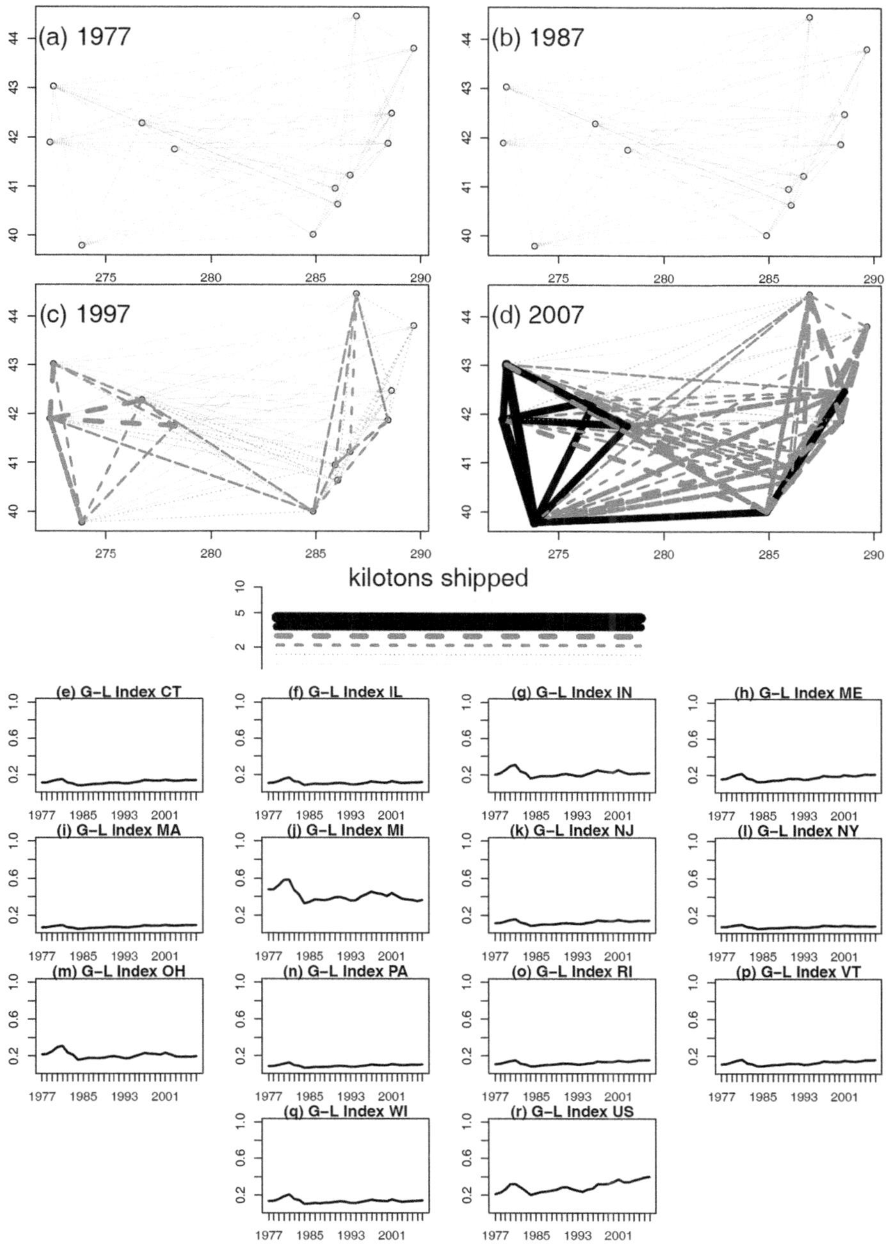

Fig. 4.2 (continued)

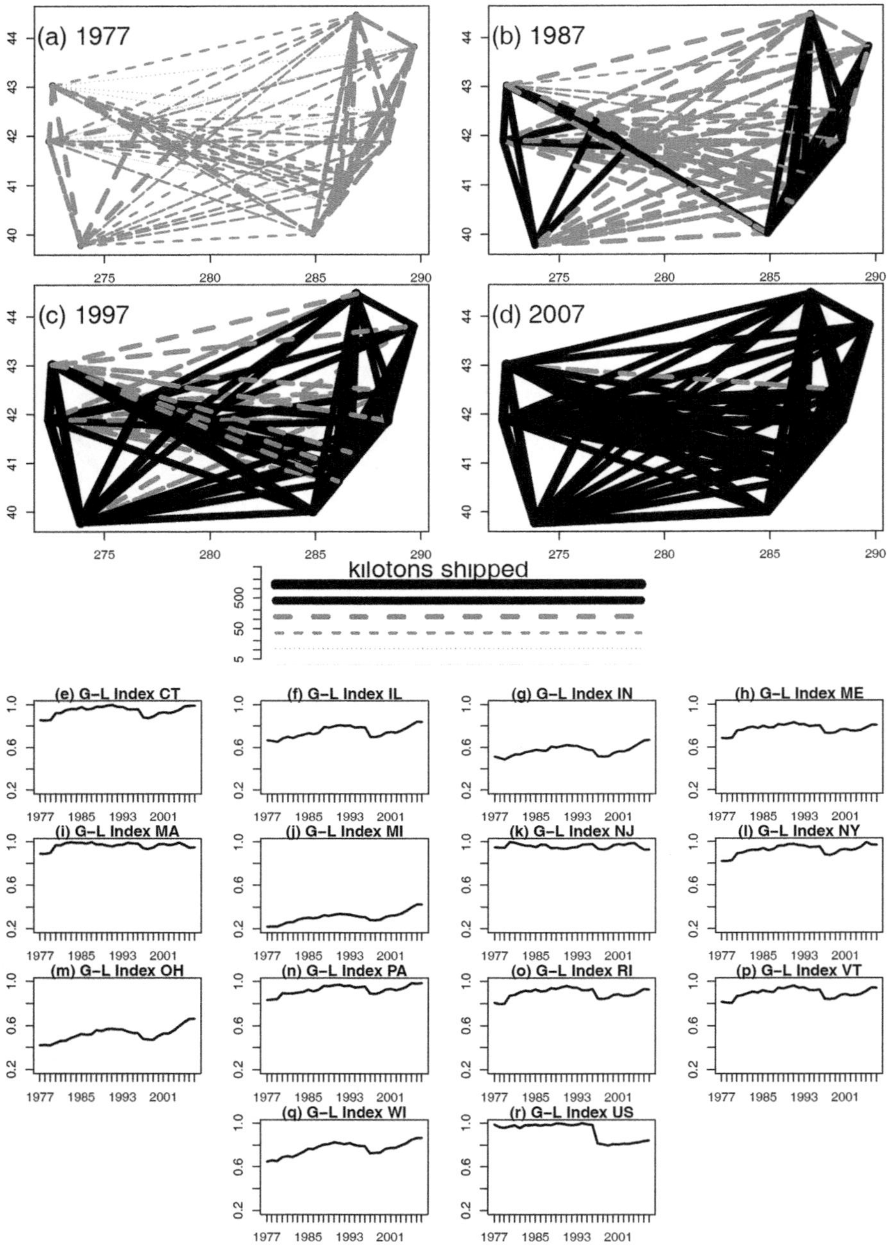

Fig. 4.2 (continued)

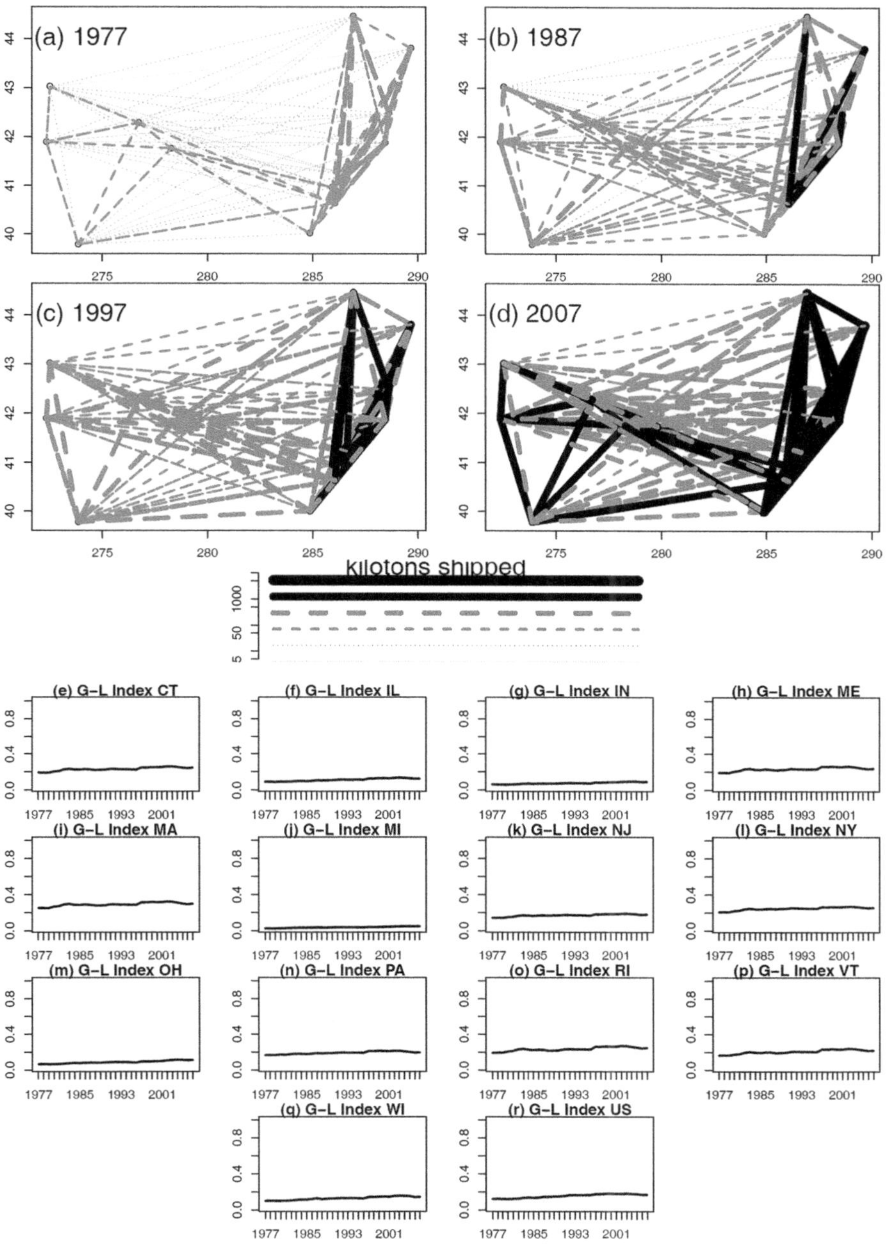

Fig. 4.2 (continued)

Table 4.3 Five-year averages of cross-section freight movement-related BC emissions from latitudes 40 N to 44 N

		1977–1982	1983–1987	1988–1992	1993–1997	1998–2002	2003–2007
latitude (degrees N)	44	1.2E + 09	1.6E + 09	1.9E + 09	2.4E + 09	2.5E + 09	2.3E + 09
	43	4.9E + 09	6.4E + 09	7.6E + 09	9.4E + 09	9.8E + 09	9.2E + 09
	42	6.5E + 09	8.6E + 09	1.0E + 10	1.3E + 10	1.3E + 10	1.2E + 10
	41	3.3E + 09	4.3E + 09	5.2E + 09	6.4E + 09	6.7E + 09	6.4E + 09
	40	1.4E + 09	1.9E + 09	2.2E + 09	2.7E + 09	2.8E + 09	2.7E + 09

Units: molecules/cm^2/second

4.2 Nonpoint Source Emissions Associated with Freight Movement

As noted above, a major concern raised by the exponential increase in freight movement is that it will, if it has not already done so, have a deleterious effect on air quality in major transportation corridors. Table 4.3 summarizes preliminary estimates of black carbon (BC) emissions associated with commodity shipping from all sectors in the 13 states and rest of the United States here studied by heavyduty diesel trucks (HDDV) and railroads across a latitudinal band crossing some east-west shipping lanes (see Fig. 4.1). This particular latitudinal band is meant to represent the shipment trends observed in the rest of the domain. These emission estimates, obtained by employing coefficients from the EPA MOVES (Motor Vehicle Emissions Simulator) software with data generated in this study, are stated in units of molecules of BC per centimeter squared per second, as required by the Community Earth System Global Climate Model (CESM).

These emission estimates were created by dividing the domain (see Fig. 4.1) into 0.5 × 0.5 degree latitude and longitude grid cells and summing the emissions along each link (for interstate shipping) and for each state (for intrastate emissions) distributed by county population.

Note that the northernmost and southernmost grid cells have comparable emissions, since both are outside of, or on the outskirts of, the transportation corridor. In contrast, the innermost grid cell has peak emissions, which are roughly three to four times as high as the northernmost and southernmost grid cells. The innermost grid cells have multiple transportation corridors within their boundary. This spatial distribution remains consistent throughout the time series.

Also, note that every grid cell has a minimum value at the beginning of the time series (1977–1982). These values roughly double by the 1993–1997 timeframe, after which they plateau, rising much more gradually and eventually decreasing by the end of the time series (2003–2007). This plateau is a combination of (1) increasing transportation throughout the time series and (2) rapid and widespread emission regulations and technological advancements in the 1990s and 2000s leading to emissions reductions, as one might expect from US DOT (2003) and Tao et al. (2007).

4.3 Conclusions and Continuations

We have argued in Chaps. 2 and 3 that if one wants to understand the impacts of globalization on economic geography and the physical environment and confront theoretical explanations of these developments empirically, then there is a need to produce time-series data on interregional/interstate inter-industry sales and their associated freight movements and nonpoint source emissions. We need to do so to understand better developments in a globalizing world, to improve our theorizing and forecasting abilities regarding these developments, and to support planning for further changes in economic geography and air quality, *inter alia.* In Chap. 3, we presented a methodology for generating interregional/interstate inter-industry sales. In the present chapter, we have demonstrated an application of this methodology in producing annual time series on interstate inter-industry freight flows in the Midwest–Northeast transportation corridors of the United States, and illustrated how these data can be used to produce estimates of associated nonpoint source emissions. In Chap. 5, we analyze in finer detail the relationship between the spatial distribution of black carbon emissions and their circulation through the Midwest and Northeast regions of the United States and the offsetting developments of exponentially increasing freight movement and emissions-reducing technologies.

Appendix

Following Munroe et al. (2007), for a single industry, *i*, in some state, *s*, the Grubel–Lloyd index of intra-industry trade is computed as follows:

$$\begin{aligned}
\mathrm{SIIT}_{si} &= 1 - \left\{|X^e_{si} - M^e_{si}| \div (X^e_{si} + M^e_{si})\right\},\\
\text{where } X^e_{si} &= X_{si}\{(X_s + M_s)/2X_s\},\\
M^e_{si} &= M_{si}\{(X_s + M_s)/2M_s\},\\
X_{si} &= \text{exports of industry } i \text{ in state } s,\\
M_{si} &= \text{imports of industry } i \text{ in state } s,\\
X_s &= \text{total exports by all industries in state } s,\\
M_s &= \text{total imports by all industries in state } s.
\end{aligned}$$

For the aggregate economy (all industries) of a state, the index would be computed as follows:

$$\mathrm{SIIT}_s = 1 - \left\{\left[\sum_i |X^e_{si} - M^e_{si}|\right] \div \left[\sum_i (X^e_{si} + M^e_{si})\right]\right\}.$$

A value of 1.0 indicates perfect trade overlap, whereas a value of zero indicates perfect specialization.

References

Boyce D (2002) Combined model of interregional commodity flows on a transportation network. In: GJD H, Sonis M, Boyce D (eds) Trade, networks, and hierarchies: modeling regional and interregional economies. Springer, Berlin, pp 29–40

Chen J, Donaghy KP (2012) Econometric estimation and qualitative analysis of a dynamic commodity-flow model for studying impacts of globalization and infrastructure system changes on regional air quality. Paper presented at the North American Meetings of the Regional Science Association International, 8 November, Ottawa, Canada

Grubel HG, Lloyd PJ (1975) Intra-industry trade. Macmillan, London

Ham H, Kim TJ, Boyce D (2005) Implementation and estimation of a combined model of interregional multimodal commodity shipments and transportation network flows. Transp Res B 39:65–79

Hewings GJD, Parr JB (2009) The changing structure of trade and interdependence in a mature economy: the US Midwest. In: McCann P (ed) Technological change and mature industrial regions: firms, knowledge, and policy. Elgar, Cheltenham, pp 64–84

Israilevich P, Hewings GJD, Sonis M, Schindler GR (1997) Forecasting structural change with a regional econometric input–output model. J Reg Sci 37:565–590

Munroe D, Hewings GJD, Guo D (2007) The role of intra-industry trade in inter-regional trade in the Mid-West of the US. In: Cooper RJ, Donaghy KP, Hewings GJD (eds) Globalization and regional economic modeling. Springer, Heidelberg, pp 87–105

Tao Z, Williams A, Donaghy K, Hewings G (2007) A socio-economic method for estimating future air pollution emissions—a Chicago case study. Atmos Environ 41:5398–5409

United States Department of Transportation (US DOT) (2003) Modeling of advanced technology vehicles. US DOT, Washington, DC

Chapter 5
Black Carbon Emissions from Trucks and Trains in the Midwestern and Northeastern United States, 1977–2007

Abstract This chapter presents a framework for estimating black carbon (BC) emissions from heavy-duty diesel vehicles (HDDV) and trains engaged in transporting freight in the Midwestern and Northeastern United States between 1977 and 2007. The estimates produced are comparable to other existing emissions inventories. This framework is employed in attempting to answer two questions: (1) What were the trends in BC emissions from HDDV and rail transportation sources over this period and what were the major factors that drove these trends? (2) What economic sectors dominated BC emissions and what major changes in sectoral behavior occurred over this period? The framework presented allows for the direct estimation of future BC emissions under a variety of economic, technological, and regulatory scenarios through changes in transportation patterns and emission factors.

5.1 Introduction

Black Carbon (BC) aerosols influence both air quality and the global climate; they are a human health hazard (e.g., Janssen et al. 2012) with a net radiative forcing on the order of 1 W m^{-2} of which nearly one-third is attributed to fossil fuel combustion (Bond et al. 2013). In industrialized nations, including the United States, where biomass combustion (i.e., cooking and heating stoves) emissions are relatively low, on-road heavy-duty diesel vehicles (HDDVs) in the transportation sector are the dominant source of BC emissions (Bond et al. 2013). As noted above, the transportation sector in the United States has transformed dramatically since the 1970s: freight volumes have nearly tripled, increasing at a rate that is faster than the growth of US Gross National Product (GNP) (US DOT 2013); emission factors for BC from HDDVs, which account for over 50% of US BC transportation emissions (US EPA 2012a), have declined by nearly 80% (US EPA 2012a, b); and the transportation

The principal author and lead researcher for this chapter was Benjamin Brown-Steiner. Peter Hess provided valuable input.

K. Donaghy et al., *The Co-evolution of Commodity Flows, Economic Geography, and Emissions*, Advances in Spatial Science,
https://doi.org/10.1007/978-3-030-78555-0_5

sector has transformed as a part of an increasingly interconnected global economy that is dependent on just-in-time deliveries of both intermediate and finished commodities (Donaghy 2012).

The ultimate impact of an increased demand for transportation and a concurrent decrease in BC emission factors is complicated. This chapter examines the factors by which freight flows impact BC emissions in the Midwestern and Northeastern United States (MNUS) over a historical period (1977–2007) in order to determine and describe the causes of the overall trend. We use time-series data on commodity flows in value terms, whose generation was discussed in Chap. 3, which we convert to freight flows in weight terms. Then, distributing the derived freight flows by mode of transport in a stylized transportation network, and employing available historical emission factors for HDDVs and rail, we estimate gridded BC emissions. Our methodology allows us to isolate the BC emissions that result from the individual influences of economic and regulatory forces. This framework provides us with the capability to examine BC emissions under a variety of economic and regulatory scenarios.

There have been many factors influencing BC emissions from freight transportation in the MNUS from 1977 to 2007. The majority of BC emissions from the transportation sector come from the HDDV fleet with a minor contribution from the rail fleet (ICF 2005). The transportation of freight by HDDVs and rail is controlled by demand from producers and consumers of finished and semi-finished products, which themselves are driven by underlying dynamics of both regional and global economies including growth, supply and demand, and globalization. At the same time, particulate matter (PM) emissions, including BC emissions, have been subject to increasingly stringent regulatory efforts from the US EPA (US EPA 2012a) and from local and regional municipalities (e.g., NYSDEC 2014).

From 1977 to 2007 the value of all freight movement in the United States grew at an annual rate of roughly 5%, which is more than twice as fast as the growth of the country's Gross National Product (GNP) (US DOT 2013). Much of the increase in demand for freight shipment was the result of aggregated economic growth as well as changes in the structure, connectivity, and infrastructure of industrial producers, which is often referred to as the geography of production or economic geography (e.g., Feenstra 1998; Donaghy 2007). In addition, the transportation sector experienced advancements in information technologies, which increased the efficiency and reliability of the freight movement. These technological advancements facilitated a transition to "just-in-time" inventory management systems where products and materials can be shipped quickly in an "on-demand" basis (e.g., Krishnamarthy 2007) with a growing trend toward regular shipments of unfinished goods between production centers throughout the production process (Castells 2000). Subsequently, the industrial production process grew more dependent upon the transportation sector. Essentially, industries have taken advantage of regional economies of scale and economies of scope (Feenstra 1998; Jones and Kierzkowski 2001), which has resulted in a "hollowing out" (or a reduction of local purchasing by firms) and "clustering" (or agglomeration of similar types of activities) of production and transportation nodes (Munroe and Hewings 2007; Hewings and Parr 2009) as industries move their production processes to locations that maximize advantages from these economies. This process resulted in factories and production centers that may have been previously distributed throughout the region becoming increasingly located in and around the most efficient transportation nodes (i.e., urban centers). An example is the regular transportation of unfinished automobile parts between Canada and the United States (Anderson and Coates 2010), where car parts experience multiple border crossings throughout the production process (Anderson 2012). This development constitutes a significant change from the traditional vertically integrated or assembly line production process (Feenstra 1998).

At the same time, the US EPA and regional municipalities became increasingly aware of the negative health effects of BC emissions (e.g., Anenberg et al. 2012) and indirectly promulgated increasingly stringent controls on BC by continuously tightening the National Ambient Air Quality Standard (NAAQS) for all particulate matter (PM) (US EPA 2012a, b, 2013). This regulatory tightening, combined with a continued increase in technological efficiencies in transportation technologies, resulted in the HDDV emission factor for BC to drop from 1.29 μg/g of fuel in 1977 to 0.39 μg/g in 2007 (from the US EPA MOVES software, US EPA 2012b). This downward trend in BC emission standards is expected to continue into the future. Data on the emission factor of BC from the private rail industry are limited. In the remainder of this chapter, we focus primarily on the emission factor per ton-kilometer as we distribute BC emissions along transportation links per ton of freight shipped.

This study examines the connection between regional BC emissions and regional economic activities through model representations of the individual behavior of sectors within the MNUS. The MNUS has a large population clustered in urban centers with high manufacturing and transportation volumes. We focus on a small enough number of states to keep the computational demands manageable. This regional focus also allows for a more complete representation of the regional activities that other inventories with global spatial coverage are often unable to capture. Instead of distributing emissions based on a population proxy, as is common practice in a number of standard emission inventories (e.g., Bond et al. 2004, 2007; Lamarque et al. 2010) we distribute emissions based on a stylized transportation network that allows us to examine subregional impacts of the transportation sector on BC air quality. We compare our methodology to other existing BC emissions inventories and explore the strengths and weaknesses of our BC emissions. The framework created through the development of the REIM and in this work allows for the direct estimation of future BC emissions under a variety of economic, technological, and regulatory scenarios through changes in transportation patterns and emission factors. (See the simulations in Chap. 8.)

We attempt in this chapter to answer two primary questions: (1) What was the trend in BC emissions from HDDV and rail transportation sources between 1977 and 2007 and what were the major factors that drove that trend? (2) What economic sectors dominated these BC emissions and what major changes occurred from 1977 to 2007? Our analysis focuses primarily on the magnitude and spatial distribution of BC emissions over the historical period rather than specific details of the temporal trend. Section 5.2 describes the methodology of this study, including details of the conversion of commodity-flow data to freight-flow data, and the process by which freight flows were distributed by mode and transport network, and the format and caveats about calculations of the resultant BC emissions. Section 5.3 evaluates these BC emissions against existing BC emission inventories. Section 5.4 explores the results of these BC emissions including the interaction between increasing freight volume and decreasing emission factors, the major contributing sectors and their trends, and the regional impact of BC emissions in the Midwestern and Northeastern United States. Section 5.5 provides further discussion and conclusions.

5.2 Methods

In this section we detail the process and steps taken to estimate a time series of gridded BC emissions from both HDDV and rail ($BC_{HDDV + Rail}$) transport using economic and industry data from 1977 to 2007 (see Fig. 5.2). We are concerned in this study with the physical effects of commerce in the form of freight flows between 13 industrial sectors (Table 5.1) located in 13 states in the MNUS (Fig. 5.1) and the rest of the United States (RUS). The generation of time series on interstate interindustry commodity flows was discussed in Chap. 2. Following the approach outlined there, 2366 time-series of inter-industry sales coefficients (13 sectors × 13 sectors × 13 states plus the RUS) and an equal number of time-series of commodity flows were obtained (Steps 1 and 2 in Fig. 5.2).

The commodity flows (in terms of dollars per year) were converted to freight tonnage per year using sector-specific conversion factors (in millions of constant year-2001 US dollars per kiloton of freight) taken from the 2007 Commodity Flow Survey (US DOT 2010) (see Table 5.1) (Step 3 in Fig. 5.2). Shipments between the various sectors can be either intrastate or interstate. Intrastate shipments (within a

Table 5.1 Input–output sector description, abbreviations, commodity flow conversion units are in millions of dollars per kiloton of freight produced by a given sector

Sector	Description	Abbreviations	Commodity Flow Conversion	NAICS Code
1	Agriculture, forestry, fishing, and hunting	AGRICU	0.89	11
2	Mining	MINING	0.26	21
3	Construction	CONSTR	0.18	23
4	Food product manufacturing	FOODPM	1.40	311
5	Chemical manufacturing	CHEMMF	2.01	325
6	Primary metal manufacturing	PMETAL	1.18	331
7	Fabricated metal product manufacturing	FMETAL	0.06	332
8	Machinery manufacturing	MACHIN	9.42	333
9	Computer and electronic product manufacturing	COMPUT	22.02	334,335
10	Transportation Equipment	TREQPT	7.73	336
11	Other Non-durable Manufacturing	ONDRMF	1.56	312-316,322-324,326
12	Other Durable Manufacturing	ODRMFR	1.63	321,327,337,339
13	Transportation, Communication, Public Utilities, Services and Government Enterprises	GOVTEN	3.10	42,44,45,48,49,51-56,61,62,71,72,81

Sources: 2007 Commodity Flow Survey (US DOT 2010), and NAICS (North American Industry Classification System), http://www.census.gov/eos/www/naics/

Fig. 5.1 Study region. Black dots indicate location of each state's most populous county (US Census, 2010) and are used as the nodes for the distribution of emissions. One additional county (San Francisco County, California) (not shown) serves as the node for the Rest of the United States (RUS) region. The grey lines indicate counties within each state in the study region

single state) are spatially allocated to each county in each state based on the county population (2007 US Census, released in 2010). Interstate shipments are allocated using a link-and-node model. This model creates a single node in each state at the center of the county with the highest population (Fig. 5.1) and connects these nodes with a great-circle path. The node for the Rest of the United States (RUS) is specified as San Francisco County, California. Each individual node has a path to every other node (for a total of 105 links between 14 nodes including the RUS). We do not include changing population demographics over the time period of the study, although the population increased by roughly 10% during this historical period (Fig. 5.6). While the actual transportation system in the United States is more complicated, this set of assumptions broadly distributes transportation emissions over the MNUS. Advantages and disadvantages of our stylized transportation network, as well as the implications of the location of the RUS node are explored below.

For interstate freight the modal allocation of BC emissions (Step 4 in Fig. 5.2) distributes black carbon emissions from HDDV and rail transport ($BC_{HDDV + Rail}$) along the links connecting each state node by multiplying the length of the great-circle link by the total tons of freight shipped for each industry (ton-km). For intrastate shipments instead of great-circle links, the freight total is multiplied by the square root of each state's surface area (km) to get each state's ton-km for each sector. Overall, and especially in and around cities and along transportation corridors, the interstate emissions are the dominant source of BC emissions.

The $BC_{HDDV + Rail}$ emissions depend on how freight is shipped. Here we neglect all BC transportation sources except those from HDDV and rail sources, as surface

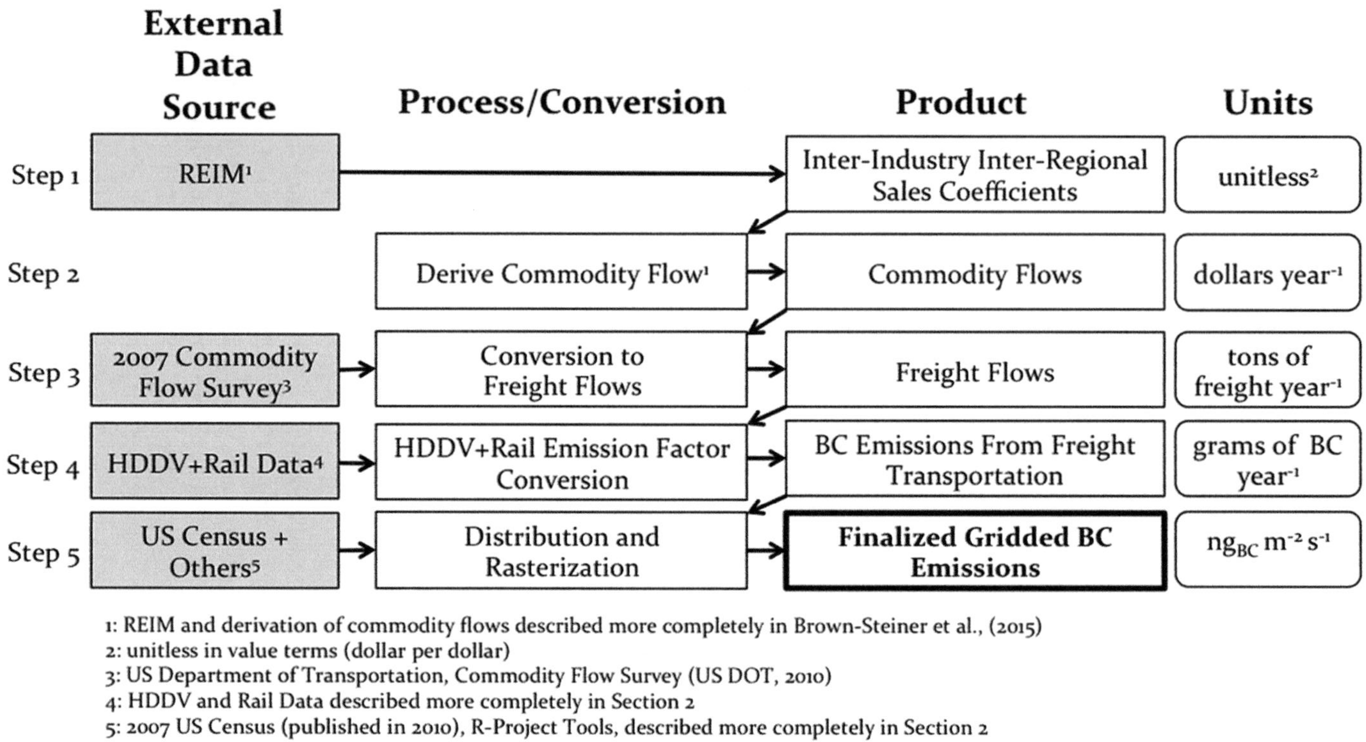

Fig. 5.2 Process schematic for this chapter. Section 5.2 describes in detail the central two columns of this figure. The leftmost column indicates which external data sources were utilized in this study. The rightmost column gives the units of the product in the third column

BC emissions are nearly all from HDDV and rail transportation modes (Uherek et al. 2010). We do not explicitly include emissions from "super-emitters" (Bond et al. 2004). A time-varying emission factor of BC from HDDVs ($EF_{BC,HDDV}$) was obtained from default MOVES simulations (in grams of BC emitted per km) (see http://www.epa.gov/otaq/models/moves/). We assumed an average of 25 tons of freight per HDDV to get an estimate of the time-varying emissions from HDDV transport in grams of BC emitted per ton-km. While HDDV particulate matter emissions standards have decreased by over 99% during the historical period (US EPA 2012b), the realized $EF_{BC,HDDV}$ has not decreased as dramatically as older model HDDVs are phased out of the active fleet. For rail, estimates of BC emissions are limited, both due to large uncertainties but also due to limited reporting from the rail industry. Rail standards for PM are on the order of 6.7–9.2 g per gallon of fuel (NESCAUM 2006) and the EPA recommends a best-guess conversion of 400 ton-miles per gallon (US EPA 2009). There are large uncertainties in the BC fraction of PM (e.g., US EPA 2012a), and we estimate a fraction or approximately 66%. We assume the emissions for rail transportation, due to the large uncertainties and lack of data availability, do not change over time. Overall our BC emissions factor from HDDVs start at 0.06 g of BC per ton-km in 1977 and decrease to 0.01 in 2007 while the BC emissions from rail are at 0.01 g of BC per ton-km throughout the historical period.

Multiplying the emission factor (grams of BC emitted per ton-km) for a particular mode of transportation (i.e., HDDV or rail) times the percent transport by that mode over the calculated ton-km for interstate and interstate shipping gives the grams of $BC_{HDDV + Rail}$ emitted (per year). In 2007, HDDVs shipped roughly 69% and rail only 15% of the total tonnage of all commodity flows, while the remainder is transported via water, pipeline, air, and unknown methods (Margreta et al. 2009). We only include HDDV and rail shipments in this study. Since the 1970s, the distribution of freight among HDDVs and rail has remained largely unchanged although the volume has increased steadily (ICF 2005). For interstate transportation, we split the total ton-miles shipped for interstate transportation from 1980 to 2007 with data from the US DOT (2015) between HDDVs and rail (for 1977–1979 we use the 1980 values) as there is preferential use of rail for heavy freight and long distances the rail mode transports more ton-miles (ICF 2005). For intrastate transportation, we give a slight preference for HDDVs (ton-miles are split 75% HDDVs and 25% rail which does not vary over time) as HDDVs are preferentially utilized for short-haul shipments (US DOT 2002). The data on HDDV and rail transportation splitting is mostly from national studies and is largely unavailable for state-by-state and commodity-by-commodity. Therefore we apply a single best-guess value over the entire region. In addition, by keeping a simplified set of parameters we will be able to create future scenarios without having to create time-varying parameter estimates for each state, sector, mode, and year.

We next distribute both intrastate and interstate $BC_{HDDV + Rail}$ emissions (grams of BC per year) by rasterizing data on each type of emission individually onto a $0.5° \times 0.5°$ grid (to produce units of grams of BC cm^{-2} s^{-1}) and sum the two together to create gridded emissions (Step 5 in Fig. 5.2). Intrastate $BC_{HDDV + Rail}$ emissions are distributed across all counties in a given state using county-level population data as a proxy. Interstate $BC_{HDDV + Rail}$ emissions are allocated uniformly onto every grid cell that touches the great-circle freight flow paths connecting each node. This process utilized the following tools from the R-Project (http://www.R-Project.org/): maptools (Bivand and Lewin-Koh 2013), raster (Hijmans and van Etten 2013), ncdf (Pierce 2014), sp (Pebesma and Bivand 2005; Bivand et al. 2008), and classInt (Bivand 2012). An evaluation of our data with the MACCity emission inventory is given in Sect. 5.3.

In order to conduct a preliminary analysis of inter-regional and inter-industry emissions from freight transportation and since this project is limited in spatial scope compared to global emissions inventories, the method described above has several limitations. First, our focus on 13 Midwestern and Northeastern states and the rest of the United States neglects freight transportation to and from Canada, which has traditionally been the largest destination for US exports and source of US imports until China recently topped Canada in US imports (Anderson and Coates 2010). For example, automobiles and automobile parts cross the United States–Canada border between the State of Michigan and the Province of Ontario multiple times during the production cycle (Anderson 2012). We do not include these in our study, as the data on the United States–Canada transportation is not available in the BEA and BTS framework. Second, while choosing San Francisco County as the node for the RUS region adequately represents the east-west transportation corridors in the United States, it neglects much of the north-south transportation corridors between the Midwest/Northeast region and the South/Southeast region (see Chun et al. 2012). Thus, any freight transportation emissions between these two regions are instead distributed between the Midwest/Northeast and the West Coast, which is likely to lead to an overestimate of BC emissions in the region. Third, the large temporal and spatial differences in many of the parameters used in our conversion process (e.g., regional distributions of HDDV and rail technologies and infrastructure) would necessitate a much more thorough analysis, but inevitably choices have to be made whether to focus on temporal resolution, spatial resolution, or more sophisticated structural representation of the underlying system. For instance, the Task Force on Hemispheric Transportation of Atmospheric Pollution, Version 2 (HTAP2) available at http://www.htap.org/ and Janssens-Maenhout et al. (2015) has a much higher spatial resolution and global coverage, but only produces emissions for 2 years. This work focuses on the MNUS and largely examines the historical time-changing economic drivers forcing BC emissions. Finally, our use of great-circle pathways between nodes does not reflect the actual transportation corridors in the US highway system, although it does create stylized corridors, which we inspect and analyze below. In contrast to emissions inventories that distribute emissions via a population proxy (e.g., MACCity), which tend to overestimate exposure in and around urban centers, this study distributes the majority of emissions (the interstate emissions) via a stylized great-circle link-and-node network, which may lead to underestimates of BC exposure along the real-world interstate highways.

5.3 Comparison with Standard Existing BC Emissions Inventories

In this section, we evaluate our $BC_{HDDV + Rail}$ emissions inventory projections and compare them to the MACCity emissions (available at http://accent.aero.jussieu.fr/MACC_metadata.php/) since the MACCity emissions have the most complete temporal coverage (1960–2010). The MACCity emissions are based on the ACCMIP emissions (available at http://accent.aero.jussieu.fr/ACCMIP_metadata.php/) for 1990 and 2000 and the RCP8.5 emissions for 2000 and 2010. The MACCity emissions give priority to regional emission inventories over global emissions inventories (Lamarque et al. 2010). The BC transportation emissions in ACCMIP, and thus the MACCity emissions, trace back to Bond et al. (2007), which themselves are an extension of BC inventories developed in Bond et al. (2004). The Bond et al. (2004) inventory is a global inventory using International Energy Agency (IEA) fuel consumption data and country-specific fuel, technology, and sector divisions in order to estimate BC emissions based on fuel type, and distributed emissions based on a population proxy. Thus, we expect geographic differences between our results and MACCity results. We also expect some differences in the time-series of the regionally averaged emissions (e.g., due to the United States–-Canada transportation).

The time-averaged (1977–2007) $BC_{HDDV + Rail}$ emissions from this study and MACCity are compared in Fig. 5.3. Our results are for HDDV+Rail only, while the MACCity emissions are estimates of HDDV+Rail emissions based off of the MACCity Total and Land Transportation BC emissions using available estimates from the US EPA (US EPA 2012a), which estimate that HDDV+Rail BC emissions

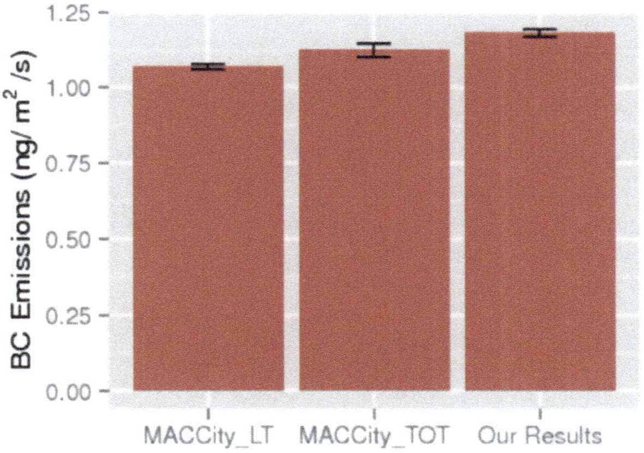

Fig. 5.3 Comparison of BC Emissions (ng/m^2/s) over this study region (Fig. 5.1) averaged from 1977 to 2007 from estimate for MACCity HDDV+Rail as 52% (US EPA 2012a) of MACCity Land Transportation (dotted red) (US EPA 2012a), estimate for MACCity HDDV+Rail as 28% (US EPA 2012a) of MACCity Total (dotted pink) (US EPA 2012a), and our results. Error bars cover the 95% confidence interval over 1977–2007

are 28% of total BC emissions and 52% of land transportation emissions (US EPA 2012a). These fractions do not vary in time and so are rough approximations of the actual proportion of HDDV+Rail BC emissions. Our emissions are of comparable magnitude as the MACCity HDDV+Rail estimate with a 10–20% high bias compared to the MACCity estimates.

Figure 5.4 compares the spatial distribution in the year 2007 of our BC emissions with the scaled estimate of the MACCity transportation BC emissions. For the MACCity BC emissions, it can be seen that the emissions are distributed largely via a population proxy as BC emissions cluster around major urban areas. Our emissions are distributed both within urbanized regions and between urban regions along idealized great-circle transportation corridors. The largest differences between our emissions and the MACCity emissions are in areas where our results have distributed BC emissions between cities and the MACCity emissions have not. The relative difference (i.e., (MACCity – This Study)/(MACCity + This Study)) between the MACCity emissions and our emissions (Fig. 5.4c with warm colors indicating the MACCity emissions are greater and cool colors indicating our

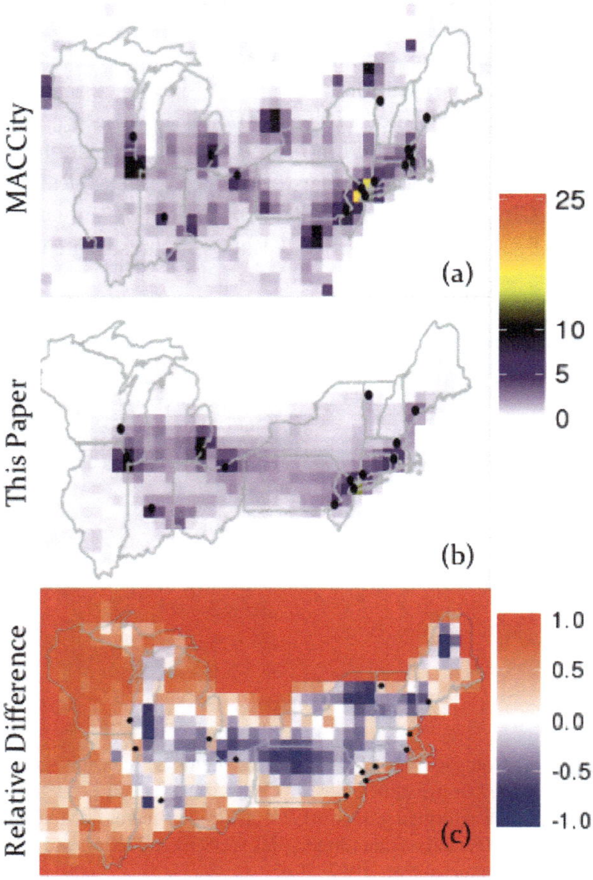

Fig. 5.4 BC transportation emissions for 2007 from: (**a**) MACCity_TOT (See Fig. 5.3); (**b**) our study (both in ng/m^2/s); and (**c**) the Relative Difference between MACCity and our results; Relative Difference is defined as (MACCity – Our Results)/(MACCity + Our Results). Black dots mark the counties of the highest population in each state (same as in Fig. 5.1) which serve as the nodes in our distribution process (Sect. 5.2). For (**c**) warm colors indicate MACCity is higher, cool colors indicate our emissions are higher. See text for descriptions of emissions sources and derivations

emissions are greater) is largest along our idealized east-west transportation corridor, where our BC emissions are ~50% higher than the MACCity emissions. This is due to our treatment of the stylized transportation network (see Sect. 5.2). Over Northern Pennsylvania, the MACCity emissions are lower than our emissions as we have a stylized transportation corridor between the East Coast and the Midwest. At the edges of our region, the relative difference between the MACCity emissions and our emissions approaches 1.0 as we do not simulate transportation patterns outside of the MNUS.

5.4 Results

Figure 5.5 gives the total $BC_{HDDV + Rail}$ transportation emissions over our region divided into 13 sectors (identified in Table 5.1) from 1977 to 2007, such that the sectors are classified according to which sector is doing the shipping (e.g., transportation emissions from a given sector to every other sector). Emissions from the actual manufacturing processes are not included in this study. These totals include interstate both intrastate and interstate transportation emissions. The dominant sector contributing to $BC_{HDDV + Rail}$ emissions is FMETAL, representing 50–70% of total BC emissions throughout the historical period. The construction (CONSTR), other nondurable manufacturing (ONDRMF), and food product manufacturing (FOODPM) sectors contribute smaller but non-negligible $BC_{HDDV + Rail}$ emissions,

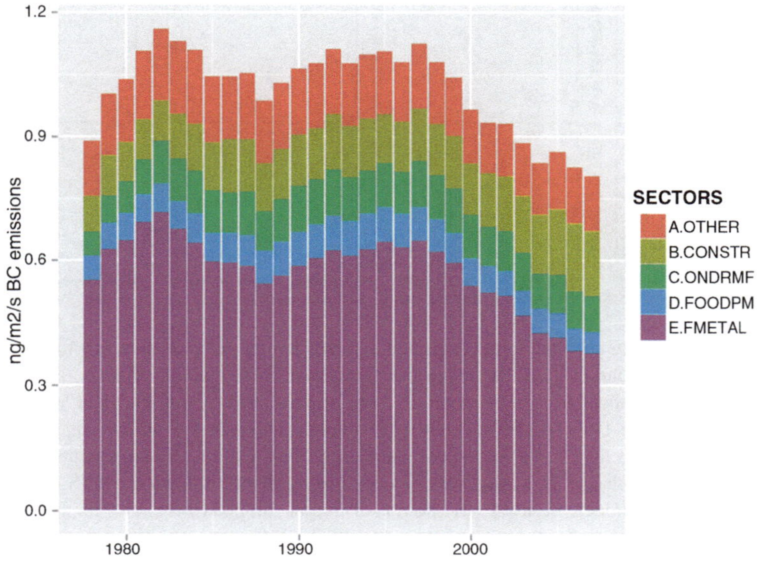

Fig. 5.5 Stacked Bar Plots of $BC_{HDDV + Rail}$ Emissions over the entire study region from 1977 to 2007 by Sector (The OTHER sector contains the sum of the nine sectors not explicitly specified in the figure)

with other sectors contributing less than 10% of the $BC_{HDDV + Rail}$ emissions. Both these FMETAL and CONSTR sectors are relatively small contributors to overall economic output (e.g., Tao et al. 2010), but both regularly transport heavy freight. BC emissions from FMETAL have two peaks in 1982 and 1997 with a steady decrease after 1997. The CONSTR sector shows little variation and only moderate growth in BC emissions throughout the historical period. We will examine individual sectors in more detail later in this section, but it is evident from Fig. 5.5 that the decline in the total BC emissions from 1997 to 2007 is largely explained by the decline in $BC_{HDDV + Rail}$ emissions from the FMETAL sector during that period.

An understanding of the historical BC emission trend requires an understanding of three underlying trends: population growth, GDP growth, and freight dynamics (in both tons and ton-miles of freight shipped). Figure 5.6 plots these together both in real terms (top) and in the percent change per year (bottom) for the 13 states in our region. Between 1977 and 2007, the population, chained GDP (with the effects of inflation removed), tons shipped, and ton-miles of freight all increased although the growth in total transported tonnage (Fig. 5.6b) is nearly double the growth in real GDP over the same period (Fig. 5.6a). The growth in total ton-miles fluctuated, sometimes growing at the same pace as the total tonnage shipped and sometimes growing at a smaller pace, or not growing at all (Fig. 5.6d). We will further analyze these factors when we examine trends in individual sectors below.

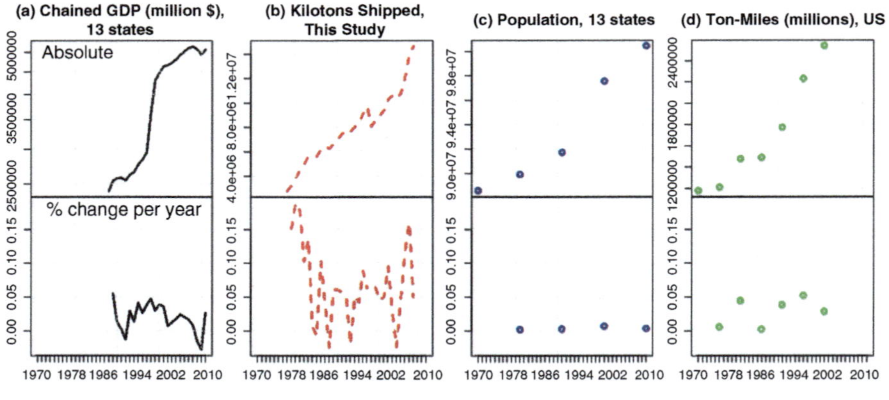

Fig. 5.6 Time series in absolute terms (top) and percent change per year (bottom) for (**a**) chained GDP (millions of dollars), which removes the growth in GDP due to inflation, and thus is a representation of the absolute growth in GDP, from http://www.bea.gov for the 13 states in this study. Note that the pre-1997 GDP is in chained 1997 dollars and post-1997 GDP is in chained 2009 dollars. GDP in 1997 is the average of the two databases. Note that the jump in chained GDP in the year 1997 is due to a change in methodology in calculating this term from the BTS database (see www.bea.gov/regional/docs/product); (**b**) kilotons of freight shipped (this study) for all regions; (**c**) population for the 13 states in our region (from the 2007 US Census, published in 2010); and (**d**) million ton-miles from HDDV and railroad transportation for the entire US ending in the year 2000 (US DOT 2005). Note the jump in chained GDP in (**a**) in the year 1997 is due to a change in methods in calculating this term from the BTS database (see http://www.bea.gov/regional/docs/product/)

The methodology utilized to produce the region's $BC_{HDDV + Rail}$ emissions (Sect. 5.2) allows us to separate the competing influences of freight growth and decreasing EF_{BC}. We can keep the EF_{BC} constant at 1977 levels while allowing the transportation changes to occur to produce one set of BC emissions (EF^{const}). We can also keep the transportation constant at 1977 levels while allowing the EF_{BC} changes to proceed as they have historically (TR^{const}). Under the TR^{const} case, the distribution of freight transportation between HDDV and rail is allowed to match historical levels, which results in slight increases in BC emissions during the beginning of the TR^{const} time series. By comparing these two trends we can explore the underlying causes of the changes in $BC_{HDDV + Rail}$ emissions (Fig. 5.5).

The $BC_{HDDV + Rail}$ emissions, from both HDDVs and Rail (black), as well as the EF^{const} (red) and TR^{const} (blue) emissions are plotted together in Fig. 5.7 for four

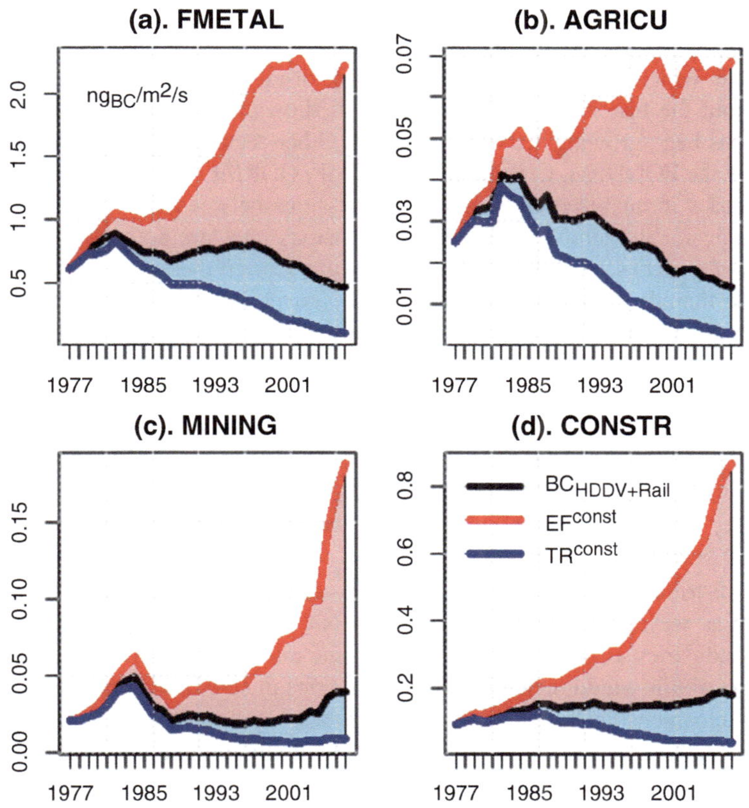

Fig. 5.7 Historical time series of $BC_{HDDV + Rail}$ emissions for four representative sectors (in $ng_{BC}/m^2/s$) in the study region (black), plus theoretical emissions if EF_{BC} is held constant at 1977 levels (EF^{const}, red) and if shipping/transportation volumes were held constant at 1977 levels (TR^{const}, blue). Note that the vertical axis is not the same for each sector. The four sectors represent different trends in temporal evolution of sectors including (**a**) and (**b**) growth and plateau/decline, (**c**) continued growth, and (**d**) varied growth

sectors (FMETAL, AGRICU, CONSTR, and MINING). Inspection of individual sectors in Fig. 5.7 show dramatic differences in both magnitude and temporal trend between sectors, and also allows us to explain some of the trends in total BC emissions from Fig. 5.6. We can classify the temporal trends of individual sectors into three categories based on the evolution of their freight volumes. The evolution of freight volumes can be ascertained from the temporal emissions of BC assuming no change in emission factor (EF^{const}, red line in Fig. 5.7). The first category, represented by the FMETAL sector in Fig. 5.7a, is *growth and plateau/decline*, in which continuous growth of freight volumes is noted until the mid-1990s after which freight volumes either level off or decline. Other sectors (not shown) that demonstrate similar temporal evolution are the FOODPM, COMPUT, and TREQPT sectors. The second category, represented in Fig. 5.7b by the AGRICU sector and Fig. 5.7c by the MINING sector, includes sectors in which freight flows have increased and decreased throughout the historic period, which we call *varying growth and decline.* Other sectors that follow this trend are the PMETAL and MACHIN sectors. The third category contains all sectors that show *continued growth*, in which freight flows grow with little or no variation in the rate of growth throughout the historical period. Figure 5.7d shows the CONSTR sector growth, which has been growing steadily over time. Other sectors that show this temporal trend are the FOODPM, CHEMMF, ONDRMF, ODRMFR, and GOVTEN sectors.

Increases in analyzed $BC_{HDDV + Rail}$ emissions between 1997 and 2007 when accounting for the temporal changes in emissions rates are noted for some of the *continued growth* sectors or the *varying growth sectors* if their rate of overall growth is greater than the concurrent decreases in EF_{BC} over the same period. These sectors include the CONSTR and GOVTEN sectors (both *continued growth*) and the MINING sector (with *varying growth and decline*). For all other sectors, $BC_{HDDV + Rail}$ emissions between 1977 and 2007 remained nearly unchanged (TOTAL, CHEMMF, ONDRMF, and ODRMFR) or declined (AGRICU, FOODPM, PMETAL, FMETAL, MACHIN, COMPUT, and TREQPT). The total $BC_{HDDV + Rail}$ emissions largely follow the FMETAL trend, although a higher variability is noted in the overall trend along with a greater decline in BC emissions after 1997.

The distribution of $BC_{HDDV + Rail}$ emissions is not expected to be spatially uniform as we expect to see a "hollowing out" (Munroe and Hewings 2007) of certain industries as production centers become more centralized (Donaghy 2012). Evidence of this would include shifts in emissions in certain sectors to certain areas or a notable intensification of emissions to developing production centers. Figure 5.8 gives $BC_{HDDV + Rail}$ emissions averaged into a set of subregions for the MACCity inventory and our results. Descriptions of the subregions are provided in the caption for Fig. 5.8, which provides values in absolute terms (Fig. 5.8a) and scaled to the maximum value (Fig. 5.8b). Both inventories show BC emissions are highest in and around urban centers, in part because production centers are concentrated in these

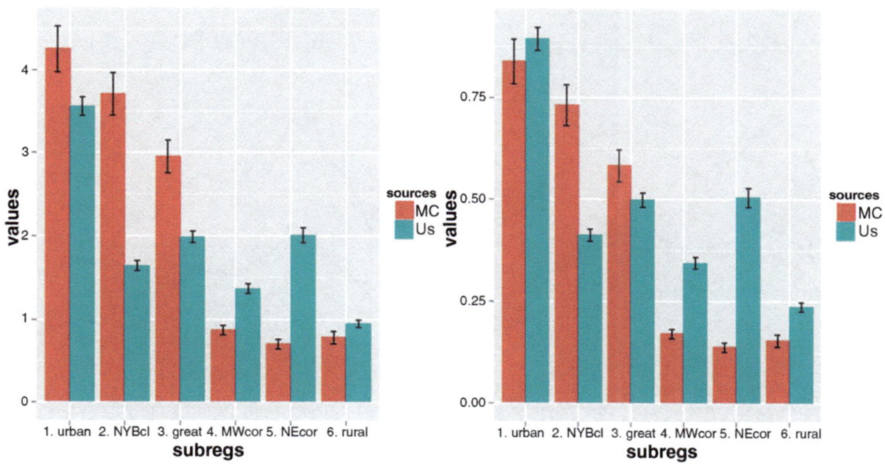

Fig. 5.8 $BC_{HDDV + Rail}$ Emissions from various subregions for MACCity (red) and our results (blue) for absolute values (left) and for values normalized to the maximum value (right). MACCity values are estimated as 28% of total anthropogenic emissions. Subregions are defined as follows: urban is an average of 6 major cities (New York City, Boston, Cleveland, Detroit, Chicago, and Indianapolis); NYBcl is the grid cells between the greater New York City and Boston regions but not the cities themselves; great is the greater urban areas, defined as the grid cells adjacent to (and including) the 6 major cities; MWCor and NEcor are the Midwestern Corridor (between Cleveland/Detroit and Chicago) and the Northeastern Corridor (between New York City/Boston and Cleveland), respectfully; and rural is a selection of rural grid cells in northern New York

regions but also because both inventories, to varying degrees, use population as a proxy for the distribution of emissions. While MACCity has similar BC emissions in all nonurban subregions (the MW and NE Corridors and the Rural Subregion), our results show a greater differentiation among subregions.

An examination of the regional differentiation of emissions from individual sectors, showing overall changes in $BC_{HDDV + Rail}$ is provided in Fig. 5.9. For both FMETAL and AGRICU (Fig. 5.9a, b), which represent the *growth* and *plateau/decline* category, the modest growth of freight volumes is insufficient to create increasing BC emissions, even in the major cities where our framework has the most growth. In contrast, the dramatic growth in freight volumes from the CONSTR sector, representative of the *continued growth* category causes increased BC emissions despite reductions in the EF_{BC} (Fig. 5.9c). Finally, the MINING sector (Fig. 5.9d), which represents the *varied growth* category, has two similar peaks in $BC_{HDDV + Rail}$ in 1982 and 2006, but we can see that the first peak is a result of modest growth in freight volumes at a time when EF_{BC} was relatively high, while the second peak is a result of very rapid growth in freight volumes at a time when EF_{BC} was relatively low.

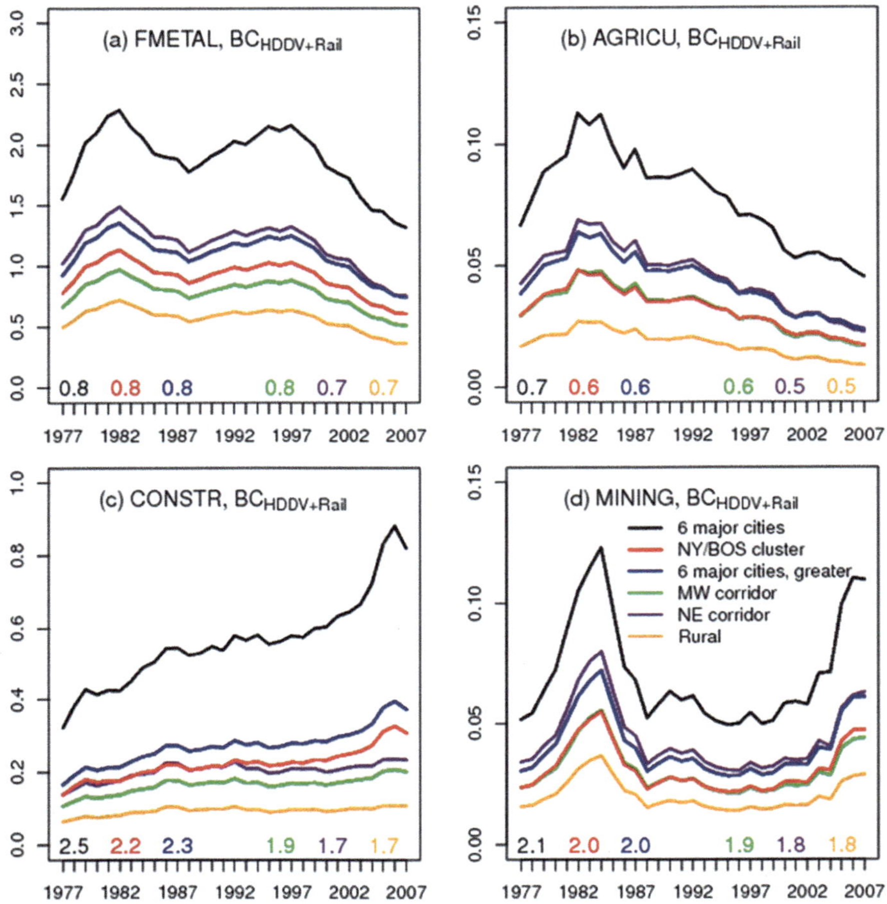

Fig. 5.9 $BC_{HDDV + Rail}$ Emissions ($ng_{BC}/m^2/s$) from various sub-regions (descriptions in Fig. 4.8) for: FMETAL (**a**), AGRICU (**c**), CONSTR (**e**), and MINING (**g**) and BC emissions under the EF^{const} case for: FMETAL (**b**), AGRICU (**d**), CONSTR (**f**), and MINING (**h**). Sectors chosen as representatives of emission trend categories: FMETAL and AGRICU represent the *growth and plateau/decline* category, CONSTR represents the *continued growth* category, and MINING represents the *varied growth* category

5.5 Discussion and Conclusion

In this chapter, we have established a framework to estimate BC emissions from heavy-duty diesel vehicles (HDDVs) and rail transportation in the Midwestern and Northeastern United States (MNUS) from 1977 to 2007 using available economic and derived shipment data. This work builds upon Donaghy and Chen (2011), Chap. 3, and Brown-Steiner et al. (2015), Chap. 4, in which a regional econometric input–output model (REIM) is developed, time series of freight flows derived, and

BC transportation emissions ($BC_{HDDV + Rail}$) are distributed and gridded. This modeling approach allows the parsing of transportation emissions of BC by state and by sector, which in turn allows for a more detailed analysis of the overall trends in $BC_{HDDV + Rail}$ emissions. We isolate the influence on $BC_{HDDV + Rail}$ emissions of changes in the economic sector and the resulting change in demand for freight shipments and the influence of changes in the EF_{BC} due to increased regulatory efforts and technological and efficiency innovations. The $BC_{HDDV + Rail}$ emissions derived from our framework are comparable to other existing BC emissions inventories (Sect. 5.3). Note that throughout this section, unless otherwise specified, we refer to the total emissions ($BC_{HDDV + Rail}$) and not emissions under the cases of constant BC emission factor (EF^{const}) or constant 1977 freight transportation (TR^{const}).

While our analysis highlights many of the complex changes impacting $BC_{HDDV + Rail}$ emissions from transportation sources from 1977 to 2007, the scope of this study has required externalizing and simplifying many potentially important factors. Subsequently, much of our analysis has focused on the magnitude of BC emissions over our historical period rather than the temporal trend of the emissions as there are large uncertainties and a lack of year-to-year data availability of many of the estimated parameters. In addition, we do not simulate international transportation here; we only simulate transportation between industries in the United States while focusing on a limited number of states. Indeed by choosing to include only 13 states in the MNUS and using San Francisco, California as our node to represent the rest of the RUS we do not simulate transportation between the MNUS and the Southeastern United States. We also simplify changes in the rail transportation fleet, in part since information on many of these changes is not publically available and remains highly uncertain. We also ignore the growth of intermodal transportation, which is where freight is transferred from one mode to another during the shipping process (Costello 2013), or changes in HDDV and rail competition, which increased after the industry deregulation of the Motor Carrier Act of 1980 (Keebler 2002; Costello 2013). We do not include off-road emissions as the REIM does not estimate off-road emissions.

Here we ask again the two questions posed at the beginning of this paper: what are the changes in BC emissions between 1977 and 2007, and what economic sectors dominate the BC emissions? For our first question, our results show that $BC_{HDDV + Rail}$ emissions from freight transportation has overall exhibited little trend over the past 30 years, with the 2007 emissions only 10% lower than the 1977 emissions. However, we do see a peak in the early 1980s, when emissions were approximately 40% higher than the 2007 values, and another peak during the 1990s, when emissions were approximately 25% higher than the 2007 values, followed by a gradual but steady decline from 1997 to 2007. These results are comparable to the MACCity inventory for the dates in which they overlap, although our results have annual variability while MACCity has only decadal variability (Fig. 5.3). The framework of this study allows us to explore in depth the concept that this overall trend in total BC from transportation emissions is a result of two competing and nearly counterbalancing factors: freight transportation has increased by a factor of around 5 while EF_{BC} has decreased by nearly 80% over the same historical time period. Without the increasingly stringent EF_{BC} regulations total BC emissions from transportation would be nearly five times as high as current levels (Fig. 5.7). Inversely, without the increased demand for freight transportation, $BC_{HDDV + Rail}$ emissions could be roughly one-quarter of current levels. The growth of freight transportation volume is not uniform among sectors, which show a variety of temporal changes over the historical period. In contrast, the change in EF_{BC} is constant across sectors and states in our framework. Taken together, the decrease in the EF_{BC} has enabled increases in freight transportation volumes without resulting in increased BC emissions (e.g., US EPA 2012a).

For our second question, we find that throughout the historical period one industrial sector has dominated the $BC_{HDDV + Rail}$ emissions signal in the MNUS: fabricated metal (FMETAL). The FMETAL sector includes any industrial process, which makes metal objects or performs finishing operations on metal objects. This sector, which emitted over half of the $BC_{HDDV + Rail}$ emissions in 1977 peaked in 1982 and 1996 but with emissions gradually decreasing from 1997 to 2007, represents roughly one-third of total BC transportation emissions in 2007 (Fig. 5.5). Tao et al. (2010), looking only at the Midwestern United States, found a comparable drop in primary and fabricated metal manufacturing and attribute the drop to globalization (i.e., production of these goods requires less sophisticated technology and is readily replaced by importation).

The other sectors, which together make up about one-third of the total BC emissions, show dramatically different trends during the historical period. Some sectors show multiple maximums and minimums throughout the historical period. For example, the transportation emissions from the MINING sector peaked in the early 1980s, decreased until 2000, and peaked again by 2007 (Fig. 5.9d). Tao et al. (2010) also note this peak in the early 1980s and attribute it to the oil price spike as well as to a decline in domestic use of coal. Between 2000 and 2007, the number of US mining employees increased by 50% (US Bureau of Economic Analysis, "Table 6.4C, D: Full Time and Part Time Employment by Industry," accessed June 3, 2015).

We also find that changes in $BC_{HDDV + Rail}$ emissions are not uniform across the MNUS. In almost every sector, $BC_{HDDV + Rail}$ emissions have increased the most in and around urban centers, which are serving as nodes in production networks and the freight distribution network. This is consistent with real-world changes to the transportation network, which we explore in detail below. Even in regions that have decreasing overall emissions, the decreases are usually smallest in and around urban centers. In contrast, BC emissions from the MINING sector increased in all subregions, with the largest increase noted in and around urban regions. Conversely, in sectors that show decreasing total $BC_{HDDV + Rail}$ emissions (such as the FMETAL and AGRICU sectors) the decreases are found largely outside of the urban centers. Large decreases are noted in the stylized transportation corridors (i.e., MW and NE Corridors) while smaller decreases are noted in rural regions.

Our results are comparable to other inventories (i.e., MACCity) for urban centers and the greater urban areas immediately surrounding urban centers, but differ in that our results show greater BC emissions along transportation corridors (Fig. 5.8) than other inventories. We ascribe the differences in subregional trends in BC emissions to regional changes in the producers of finished and semifinished goods and the freight distribution system, which connects them, all under the influence of broader scale globalization. We caution, however, that we have used a stylized transportation network that cannot capture the subtleties of the actual changes. For example, there is growing evidence that the infrastructure and networks that make up the transportation sector between production centers are changing non-uniformly (Feenstra 1998; Donaghy 2007; Bowen Jr. 2008; Rodrigue and Notteboom 2010) and that production centers are shifting from city centers to suburban centers (Cidell 2010). Many of these changes are not represented in our framework. Moreover, urban transportation trends in BC emissions are difficult to verify from atmospheric measurements. Overall, some urban centers have shown decreases in measured BC concentrations (e.g., Kirchstetter et al. 2008; US EPA 2012a), while others show little or no trend, especially within the past decade (e.g., Allen 2014). However, BC emissions in urban centers are from a variety of sources so attributing trends in atmospheric measurements to changes in transportation is difficult. For instance, decreasing BC in Boston from 2002 to 2005 is attributed to changes in the metro and school bus fleets, as well as changes to the highway infrastructure, while little trend in BC concentrations is noted from 2005 to 2012 (Allen 2014). Additionally, diesel fleet composition has changed over time, and BC air quality in cities may be impacted more by off-road diesel vehicles than on-road diesel vehicles (e.g., Kirchstetter et al. 2008). Moreover, there is high spatial variability of BC concentrations within any given city (e.g., NYC Health 2015) and each urban center contains a variety of different competing factors that can influence BC concentrations. Making direct comparisons between different urban centers is also difficult because data on traffic patterns within cities is limited (NCFRP 2013), while studies that do exist have limited time horizons (e.g., Allen 2014; NYC Health 2015).

This study highlights some of the benefits and trade-offs inherent in regional-scale emissions inventories, especially in contrast to the benefits and trade-offs inherent in global emissions inventories. For instance, our limited spatial coverage allowed for a more detailed representation of the transportation network, here a stylized link-and-node network, but future work could create emissions along actual transportation corridors. In contrast, global inventories often distribute emissions via a population proxy. In addition, by including estimates of state and sector transportation data in this study we had the structure, but not the data availability or resources, to gather and include HDDV and rail modal distributions, details of freight loads, and emission factors on a sector-by-sector, state-by-state, and year-by-year basis. This type of research would be extremely valuable in understanding differences and potential biases in global versus regional emissions inventories.

References

Allen G (2014) Analysis of spatial and temporal trends of black carbon in Boston. Prepared by the Northeast States for Coordinated Air Use Management

Anderson WP (2012) Public policy in a cross-border economic region. Int J Public Sect Manag 25:492–499

Anderson WP, Coates A (2010) Delays and uncertainty in freight movements the Canada-US border crossings. In: Transportation logistics trends and policies: successes and failures, Proceedings of the 45th annual conference of the Canadian transportation research forum, pp 129–143

Anenberg SC, Schwartz J, Shindell D, Amann M, Faluvegi G, Klimont Z, Janssens-Maenhout G, Pozzoli L, Van Dingenen R, Vignati E, Emberson L, Muller NZ, West JJ, Williams M, Demkine V, Hicks WK, Kuylenstierna J, Raes F, Ramanathan V (2012) Global air quality and health co-benefits of mitigating near-term climate change through methane and black carbon emission controls. Environ Health Perspect 120:831–839

Bivand RS (2012) ClassInt: choose univariate class intervals. R package version 0.1–19. http://cran.r-project.org/package=classInt

Bivand RS, Lewin-Koh N (2013) Maptools: tools for reading and handling spatial objects. R package version 0.8–23. http://cran.r-project.org/package=maptools

Bivand RS, Pebesma EJ, Gomez-Rubio V (2008) Applied spatial data analysis with R. Springer, New York

Bond TC, Streets DG, Yarber KF, Nelson SM, Woo J-H, Klimon Z (2004) A technology-based global inventory of black and organic carbon emissions from combustion. J Geophys Res 109: D14203

Bond TC, Bhardwaj E, Dong R, Jogani R, Jung S, Roden C, Streets DG, Trautmann NM (2007) Historical emissions of black and organic carbon aerosol from energy-related combustion, 1850-2000. Global Biogeochem Cycle 21:GB2018

Bond TC, Doherty SJ, Fahey DW, Forster PM, Berntsen T, DeAngelo BJ, Flanner MG, Ghan S, Kärcher B, Koch D, Kinne S, Kondo Y, Quinn PK, Sarofim MC, Schultz MG, Schulz M, Venkataraman C, Zhang H, Zhang S, Bellouin N, Guttikunda SK, Hopke PK, Jacobson MZ, Kaiser JW, Klimont Z, Lohmann U, Schwarz JP, Shindell D, Storelvmo T, Warren SG, Zender CS (2013) Bounding the role of black carbon in the climate system: a scientific assessment. J Geophys Res Atmos 118:5380–5552

Bowen JT Jr (2008) Moving places: the geography of warehousing in the US. J Transp Geogr 16:379–387

Brown-Steiner B, Chen J, Donaghy KP (2015) The evolution of freight movement and associated non-point-source emissions in the midwest–northeast transportation corridor of the United States, 1977–2007. In: Batabyal AA, Nijkamp P (eds) The region and trade: new analytic directions. World Scientific, Singapore, pp 177–204

Castells M (2000) The rise of the network society, 2nd edn. Blackwell, Oxford

Chun Y, Kim H, Kim C (2012) Modeling interregional commodity flows with incorporating network autocorrelation in spatial interaction models: an application of the US interstate commodity flows. Comput Environ Urban Syst 36:583–591

Cidell J (2010) Concentration and decentralization: The new geography of freight distribution in US metropolitan areas. J Transp Geogr 18:363–371

Costello B (2013) The trucking industry: the lynchpin of the U.S. economy. Bus Econ 48:195–201

Donaghy KP (2007) Globalization and regional economic modeling: analytical and methodological challenges. In: Cooper R, Donaghy KP, Hewings G (eds) Globalization and regional economic modeling. Springer, Berlin, pp 1–11

Donaghy KP (2012) Urban environmental imprints after globalization. Reg Environ Chang 12:395–405

Donaghy KP, Chen J (2011) Generating spatial time series on interstate inter-industry freight flows, Paper presented at the National Urban Freight Conference, Long Beach, CA, October 12–14

Feenstra RC (1998) Integration of trade and disintegration of production in a global economy. J Econ Perspect 12(4):31–50

Hewings GJD, Parr JB (2009) The changing structure of trade and interdependence in a mature economy: the US Midwest. In: McCann P (ed) Technological change and mature industrial regions: firms, knowledge, and policy. Elgar, Cheltenham, pp 64–84

Hijmans RJ, van Etten J (2013) Raster: geographic data analysis and modeling. R package version 2.1–25. http://CRAN.R-project.org/package=raster

ICF (2005) Assessing the effects of freight movement on air quality at the national and regional level, final report. April 2005. Prepared for the US Federal Highway Administration by ICF Consulting

Janssen N, Gerlofs-Nijland AH, Miriam E, Lanki T, Salonen RO, Cassee F, Hoek G, Fischer P, Brunekreef B, Krzyzanowski M (2012) Health effects of black carbon. WHO Regional Office for Europe, Copenhagen

Janssens-Maenhout G, Crippa M, Guizzardi D, Dentener F, Muntean M, Pouliot G, Keating T, Zhang Q, Kurokawa J, Wankmüller R, Denier van der Gon H, Klimont Z, Frost G, Darras S, Koffi B (2015) HTAP_v2: a mosaic of regional and global emission gridmaps for 2008 and 2010 to study hemispheric transport of air pollution. Atmos Chem Phys Discuss 15:12867–12909

Jones RW, Kierzkowski H (2001) A framework for fragmentation. In: Arndt SW, Kierzkowski H (eds) Fragmentation: new production patterns in the world economy. Oxford University Press, New York, pp 17–34

Keebler JS (2002) Trends in the transportation of goods in the U.S. J Transp Manag Fall 2002:1–14

Kirchstetter TW, Aguiar J, Tonse S, Novakov T, Fairley D (2008) Black carbon concentrations and diesel vehicle emission factors derived from coefficient of haze measurements in California: 1967–2003. Atmos Environ 42:480–491

Krishnamarthy A (2007) From just in time manufacturing to on-demand services, in service enterprise integration. In: Hsu C (ed) Integrated service in information systems, vol 16, pp 1–37

Lamarque J-F, Bond TC, Eyring V, Granier C, Heil A, Klimont Z, Lee D, Liousse C, Mieville A, Owen B, Schultz MG, Shindell D, Smith SJ, Stehfest E, Van Aardenne J, Cooper OR, Kainuma M, Mahowald N, McConnell JR, Naik V, Riahi K, van Vuuren DP (2010) Historical (1850–2000) gridded anthropogenic and biomass burning emissions of reactive gases and aerosols: methodology and application. Atmos Chem Phys 10:7017–7039

Margreta M, Ford C, Dipo MA (2009) U.S. freight on the move: highlights from the 2007 commodity flow survey preliminary data. US Department of Transportation, pp 1–6

Munroe D, Hewings GJD (2007) The role of intraindustry trade in interregional trade in the Midwest of the U.S. In: Cooper R, Donaghy K, Hewings G (eds) Globalization and regional economic modeling. Springer, Heidelberg, pp 87–105

NCFRP (2013) Report 23: synthesis of freight research in urban transportation planning. In: Giuliano G, O'Brien T, Dablanc L, Holliday K (eds) METRANS Transportation Center Sol Price School of Public Policy. University of Southern California, Los Angeles, CA

NESCAUM (2006) Scoping study to evaluate locomotive emissions operating in new haven, Connecticut and potential control options, Final Report. June 2006. Prepared by the Northeast States for Coordinated Air Use Management

NYC Health (2015) The New York City community air survey, neighborhood air quality 2008–2013. April 2015

NYSDEC (2014) 2014 Annual monitoring network plan, New York state ambient air monitoring program. Bureau of Air Quality Surveillance, Division of Air Resources, New York State Department of Environmental Conservation

Pebesma EJ, Bivand RS (2005) Classes and methods for spatial data in R. R News 5(2) http://cran.r-project.org/doc/Rnews/

Pierce D (2014) Ncdf: interface to Unidata netCDF data files. R Package Version 1.6.7. http://cran.r-project.org/package=ncdf

Rodrigue JP, Notteboom T (2010) Comparative North American and European gateway logistics: the regionalism of freight distribution. J Transp Geogr 18:497–507

Tao Z, Hewings G, Donaghy K (2010) An economic analysis of Midwestern US criteria pollutant emissions trends from 1970 to 2000. Ecol Econ 69:1666–1674

Uherek E, Halenka T, Borken-Kleefeld J, Balkanski Y, Berntsen T, Borrego C, Gauss M, Hoor P, Juda-Rezler K, Lelieveld J (2010) Transport impacts on atmosphere and climate: land transport. Atmos Environ 44:4772–4816

US DOT (2002) Freight shipments in America: preliminary highlights from the 2002 commodity flow survey. 2002. US Department of Transportation, Bureau of Transportation Statistics

US DOT (2005) National transportation statistics 2004, Department of Transportation, Bureau of Transportation Statistics, Washington, DC, U.S. Government Printing Office, February 2005

US DOT (2010) 2007 Economic census, transportation, 2007 commodity flow survey, United States: 2007. Issued April 2010. Bureau of Transportation Statistics and US Census Bureau

US DOT (2013) Freight facts and figures 2013. Federal Highway Administration and the Bureau of Transportation Statistics

US DOT (2015) National transportation statistics, Table 1–50, Updated April, 2015. U.S. Department of Transportation, Research and Innovative Technology Administration, Bureau of Transportation Statistics. http://www.bts.gov/publications/national_transportation_statistics/

US EPA (2009) Emission factors for locomotives, technical highlights. April 2009. Office of Transportation and Air Quality, US Environmental Protection Agency, EPA-420-F-09-025

US EPA (2012a) Report to congress on black carbon. March 2012. Department of the Interior, Environment, and Related Agencies, Appropriations Act 2010. EPA-450/R-12-001

US EPA (2012b) Development of emission rates for heavy-duty vehicles in the motor vehicles emissions simulator MOVES 2010 final report. August 2012. EPA-420-B-12-049

US EPA (2013) National ambient air quality standards for particulate matter, final rule. Fed Regist 78(10):3086–3287

Chapter 6
Some Extensions to Interregional Commodity-Flow Models

Abstract Some interregional commodity-flow models, developed in the tradition of interregional input–output modeling, take on detailed characterizations of transportation networks to extend their explanatory reach, whereas others, developed in the tradition of spatial-interaction modeling, assume detailed characterizations of production. This chapter demonstrates how features of models within each of these two traditions can be integrated into two new specifications: a partial equilibrium static formulation and a dynamic formulation of production, location, and interaction. The chapter also introduces several extensions to extant commodity-flow models, including explicit treatment of trade in intermediate goods, so-called *new economic geography* behavioral foundations for production and interindustry and interregional trade, and endogenous determination of capital investment and employment. These extensions enable commodity-flow models to be used to analyze the impacts of recent structural changes.

6.1 Introduction

To this point in the book, we have derived measures of commodity flow that we have converted to measures of freight movement and associated nonpoint source emissions. But we have not formally characterized the movement of commodities. Interregional commodity-flow models have played an important role in the development of regional science in characterizing flows of goods and services within and between regional economies. They have reflected not only changing patterns of economic geography on the ground, so to speak, but also the interregional and intraregional trade and transport activity needed to bring about and sustain those patterns. As such, commodity-flow models have integrated models of production, distribution, and transportation and have moreover served, when embedded in programming models, to support forecasting activities and planning analyses of infrastructure investments or industrial and other policies.

Particular commodity-flow models have been developed to meet the needs for analysis of particular problems at particular times and the models have reflected the

K. Donaghy et al., *The Co-evolution of Commodity Flows, Economic Geography, and Emissions*, Advances in Spatial Science,
https://doi.org/10.1007/978-3-030-78555-0_6

state of theoretical model development at those times. (See Batten and Boyce (1986) for an early review of the literature and Boyce and Williams (2015) passim for a more recent discussion of "goods transportation.") One can identify several "family trees" in commodity-flow modeling. There are models that would seem to derive from interregional input–output parentage (e.g., Leontief and Strout 1963, Moses 1960, and Kohno et al. 1994), which go on to introduce more detail of transport networks, and those which derive from spatial interaction parentage (e.g., Wilson 1970, Kim et al. 1983, Batten and Boyce 1986, Boyce 2002, and Ham et al. 2005), which go on to introduce more detail about production.[1] Given the array of models that are now in the literature, what is taken to be a commodity-flow model, versus something else, may be to a certain extent in the eye of the beholder.

This chapter demonstrates how features of models in these two families can be integrated in two new specifications: a partial-equilibrium static formulation and a dynamic formulation of location, production, and interaction. It also introduces several extensions to extant commodity-flow models: explicit treatment of trade in intermediate goods, so-called *New Economic Geography* behavioral foundations for production and inter-industry and interregional trade, and endogenous determination of capital investment and employment.[2] These extensions enable commodity-flow models to be used to analyze the impacts of such structural changes associated with globalization as the fragmentation of production, the lengthening of supply chains, and the clustering of industrial activities to enable exploitation of economies of scale and scope.

6.2 Extensions to a Static Partial-Equilibrium Model

In their 1986 review of spatial interaction, transportation, and interregional commodity-flow models, Batten and Boyce demonstrate how a static model which simultaneously deals with the location of goods production and assignment of commodity flows can be derived from the spatial-interaction principles embodied in the model formulated by Wilson (1970). The solution to the model, which minimizes shipments in ton-miles (or the equivalent) that satisfy demands for intermediate and final goods, is conditioned on a formalization of entropy, to accommodate cross-hauling of goods, and allows for flow-dependent costs. Boyce (2002) would go on to modify the specification of this model and estimate its parameters statistically. Ham et al. (2005) would further exploit this model to characterize multimodal shipping patterns in the United States. There are several extensions that can be made to this class of models that would increase their flexibility and coverage of analytical problems. For one, the Leontief input–output

[1]One could also include multimodal freight network models with production detail (e.g., Southworth and Peterson 2000).

[2]Some of these extensions have been suggested in Donaghy (2009).

technology can be replaced by a technology that allows for both substitution of inputs in production, in response to changes in factor price ratios, and economies of scale. A second extension is that delivered prices of intermediate goods can be introduced both to gain insight into how shipment patterns would respond to changes in such prices and to support explicit modeling of trade in intermediate goods. We introduce both changes in this section.

We shall adopt the following notation to characterize network flows. Nodes of the network through which goods are shipped are indexed by l and m. Links joining such nodes are indexed by a and routes comprising contiguous links are indexed by r. The length of some link a connecting two nodes for some mode of transport v is denoted by d^v_a.[3] If link a is part of route r connecting nodes l and m for mode v, an indicator variable δ^{av}_{lmr} assumes the value 1.0. It is 0 otherwise. The length of a given route from some node l to another node m for mode v, D_{lmr}, is given by the sum of link distances along the route, that is,

Length of Route r from Node l to Node m for Mode v

$$D^v_{lmr} \equiv \sum_a d^v_a \delta^{av}_{lmr}, \forall l, \forall m, \forall r, \forall v. \tag{6.1}$$

Turning to quantities shipped through the network, we index sectors engaged in production in the spatial economy by i and j. We will not differentiate between different types of final demand—e.g., consumption, government spending, investment, or exports. Let X^i_l denote the total output (in dollars) of sector i produced at node l, x^{ij}_{lm} denote inter-industry sales from sector i at location l to sector j at location m, and FD^i_{lm} denote final demand at location m for sector i's product at location l. The physical flow of sector i's product from l to m along route r by mode v is h^{iv}_{lmr}. This quantity is obtained by converting the value flow along route r from dollars to tons by means of the ratio of total annual interregional economic flow to total annual physical flow, q^i_x. The total physical flow of all commodities shipped on a link a via all routes using the link is given by:

$$f^v_a \equiv \sum_i \sum_{lmr} h^{iv}_{lmr} \delta^{av}_{lmr} + TT^v_a, \forall a, \forall v, \tag{6.2}$$

in which TT_a denotes the physical flow of commodities associated with traffic.

[3]Other studies—e.g., Ham et al. (2005)—attempt to accommodate the impact on "effective" distance of a link that congestion may have, in a sense making imputed transport cost *flow-dependent*, by employing the Bureau of Public Roads link congestion function, $d_a = d^o_a\left[1 + 0.15(f_a/k_a)^4\right]$, in which d^o_a is the physical distance covered by link a, f_a is the physical flow of commodities along the link (to be defined below) and k_a is the periodic flow capacity of the link. Because of computational challenges introduced by model specification changes discussed below, we will not follow suit but will retain link capacity constraints in the formulation of the model.

Conditions that the network must satisfy at any point in time are as follows:

Material Balance Constraint

$$X_l^i = \sum_m \sum_j x_{lm}^{ij} + \sum_m FD_{lm}^i, \forall i, \forall l; \tag{6.3}$$

Conservation of Flows Constraint

$$\sum_r h_{lmr}^{iv} = \wp^{iv} \sum_j x_{lmt}^{ij} / q_x^i + \wp^{iv} FD_{lm}^i / q_x^i, \forall i, \forall l, \forall m, \forall v; \tag{6.4}$$

where $\wp^{iv}$ denotes the probability that good i will be shipped via mode v and, following Ham et al. (2005), is taken to be a random utility function of average distance $(\overline{dist_i})$ and average value $(\overline{val_i})$ of shipment of good i by available modes v—i.e.,

$$\wp^{iv} = e^{\beta_1^{iv} \overline{val_i} + \beta_2^{iv} \overline{dist_i}} / \sum_v e^{\beta_1^{iv} \overline{val_i} + \beta_2^{iv} \overline{dist_i}}, \forall v.$$

Link Capacity Constraint

$$f_a^v \le k_a^v, \forall a, \forall v; \tag{6.5}$$

Non-Negativity and Feasibility Conditions

$$f_a^v \ge 0, \forall a, \forall v; h_{lmr}^{iv} \ge 0, \forall i, \forall l, \forall m, \forall r, \forall v; x_{lm}^{ij} > 0, \forall i, \forall j, \forall l, \forall m. \tag{6.6}$$

Equation (6.3) ensures that shipments from industry i in location l do not exceed production by the industry in that location, while Eq. (6.4) reconciles physical and value flows. Inequality (6.5) imposes physical limitations on flows over links and the conditions given in (6.6) ensure that the distribution of goods throughout the network is feasible.

Because observations on FD_{lm}^i, final demand at location m for sector i*'s* goods produced at location l, are generally not available and cannot be easily derived, we shall replace the summation of FD_{lm}^i terms in Eq. (6.3) with the aggregate quantity $\overline{\text{FD}}_l^i$, which is a measure of all of industry i's product at location l that is shipped to all sources of final demand at all locations in the network. This quantity can be determined by remainder when Eq. (6.3) is rewritten as:

Modified Material Balance Equation

$$\overline{\mathrm{FD}}^i_l\left(\equiv \sum_m \mathrm{FD}^i_{lm}\right) = X^i_l - \sum_m \sum_j x^{ij}_{lm}, \forall i, \forall l. \qquad (6.3')$$

The part of industry i's product at location l that is shipped to all sources of final demand in just location m, $\overline{\mathrm{FD}}^i_{lm}$, can be approximated by employing a gravity model formulation in which $\overline{\mathrm{FD}}^i_l$, the aggregate amount of the output from industry i at location l to be distributed to all locations, is allocated to each location m according to its relative share of total industrial output and the impedance of distance:

$$\overline{\mathrm{FD}}^i_{lm} = \overline{\mathrm{FD}}^i_l \cdot \frac{\sum_j X^j_m}{\sum_m \sum_j X^j_m} \cdot D^{*-1}_{lmr}, \forall i, \forall l, \forall m. \qquad (6.7)$$

In this expression, D^{*-1}_{lmr} is the inverse of the distance of the shortest route between locations l and m by any mode v. Employing this expression, Eq. (6.4) can now be rewritten as:

Modified Conservation of Flows Constraint

$$\sum_r h^{iv}_{lmr} = \wp^{iv} \sum_j x^{ij}_{lm}/q^i_x + \wp^{iv} \overline{\mathrm{FD}}^i_{lm}/q^i_x, \forall i, \forall l, \forall m, \forall v. \qquad (6.4')$$

We assume that production of good j at location m, at time t can be represented by a two-stage production function with constant elasticity of substitution (CES) and the possibility of increasing returns to scale, hence the possibility of imperfect competition. The output of good j at location m, X^j_{mt}, is a function of aggregates of inputs i shipped from other locations l and location m, c^{ij}_{mt}, and local labor and capital in the industry, L^j_{mt} and K^j_{mt}. The production function and input aggregator functions are written as:

Production Function

$$X^j_m = \left[\sum_i \alpha^{ij}_m c^{ij-\rho^j_m}_m + \alpha^{L_j}_m L^{j-\rho^j_m}_m + \alpha^{K_j}_m K^{j-\rho^j_m}_m\right]^{-\kappa^j_m/\rho^j_m}, \forall j, \forall m, \qquad (6.8)$$

and

Input Aggregator Function

$$c^{ij}_m = \vartheta^{ij}_m \prod_l x^{ij\varepsilon^{ij}_{lm}}_{lm}, \forall i, \forall j, \forall m. \qquad (6.9)$$

In (6.8) the α_m^{ij} parameters are factor intensity parameters, ρ_m^j is the substitution parameter, and κ_m^j is the economies-of-scale (or returns-to-scale) parameter for industry j in location m. In (6.9), ϑ_m^{ij} is the scale parameter for the aggregator of inputs i in the production of good j, and the ε_{lm}^{ij} parameters are share parameters.

We know from microeconomic theory that in equilibrium the product of the Lagrange multiplier associated with a cost minimization problem and the partial derivative of X_m^j taken with respect to any input x_{lmt}^{ij} will be offset by the input price, which in this case will be the effective delivered price, $p_{lm\cdot}^i$. Assuming imperfect competition, hence, taking the Lagrange multiplier to be the average cost of production and to equal to the output price, P_m^j, divided by a mark-up, π, the equilibrium condition can be expressed as:

$$p_{lm\cdot}^i = \frac{P_m^j}{\pi} \cdot \frac{\partial X_m^j}{\partial x_{lm}^{ij}} = \frac{P_m^j}{\pi} \cdot \frac{\partial X_m^j}{\partial c_m^{ij}} \cdot \frac{\partial c_m^{ij}}{\partial x_{lm}^{ij}}. \tag{6.10}$$

The partial derivative of X_m^j with respect to x_{lm}^{ij} can be obtained by first rewriting (6.8) in mathematically equivalent form as:

$$X_m^{j-\rho_m^j/\kappa_m^j} = \left[\sum_i \alpha_m^{ij} c_m^{ij-\rho_m^j} + \alpha_m^{L_j} L_m^{j-\rho_m^j} + \alpha_m^{K_j} K_m^{j-\rho_m^j} \right], \forall j, \forall m, \tag{6.11}$$

and differentiating both sides with respect to c_m^{ij} to obtain:

$$-\rho_m^j/\kappa_m^j \cdot X_m^{j-\rho_m^j/\kappa_m^j-1} \cdot \frac{\partial X_m^j}{\partial c_m^{ij}} = -\rho_m^j \alpha_{lm}^{ij} c_m^{ij-\rho_m^j-1}, \forall i, \forall j, \forall m. \tag{6.12}$$

Isolating the partial derivative in (6.12) yields:

$$\frac{\partial X_m^j}{\partial c_m^{ij}} = \kappa_m^j \alpha_{lm}^{ij} X_m^{j\left(\rho_m^j+\kappa_m^j\right)/\kappa_m^j} c_m^{-\rho-1}, \forall i, \forall j, \forall m. \tag{6.13}$$

The partial derivative of the input aggregator function taken with respect to x_{lm}^{ij} is just:

$$\frac{\partial c_m^{ij}}{\partial x_{lm}^{ij}} = \varepsilon_{lm}^{ij} \frac{c_m^{ij}}{x_{lm}^{ij}}, \forall i, \forall j, \forall l, \forall m. \tag{6.14}$$

Substituting from (6.13) and (6.14) into (6.10) and rearranging terms, one obtains the partial-equilibrium expression for x_{lm}^{ij}, which is the input demand equation:

Input Demand Equation

$$x_{lm}^{ij} = \frac{\varepsilon_{lm}^{ij}\kappa_m^j\alpha_m^{ij}}{\pi} \cdot \frac{P_m^j}{p_{lm\cdot}^i} \cdot X_m^{j\left(\rho_m^j+\kappa_m^j\right)/\kappa_m^j} c_m^{ij-\rho_m^j}, \forall i, \forall j, \forall l, \forall m. \tag{6.15}$$

In Eq. (6.15), the *effective delivered price* at location m of input i shipped from location l, p_{lm}^i, will be a weighted average of the delivered prices of this input arriving via different modes v and routes r. That is,[4]

$$p_{lm\cdot}^i = \sum_{rv}\left[\frac{h_{lmr}^{iv}}{\sum_{rv} h_{lmr}^{iv}} \cdot p_{lmr}^{iv}\right]. \tag{6.16}$$

In this static formulation, we take industrial outputs (expected demand levels) and output (f.o.b.) prices to be exogenously given. Delivered (c.i.f.) prices (for particular routes and modes) are a function of output prices, distance of shipment, and shipping cost per ton-mile, p^v. That is,

$$p_{lmr}^{iv} = P_l^i + p^v \cdot D_{lmr}^v. \tag{6.17}$$

The optimization problem is to choose input demands x_{lm}^{ij} and shipments h_{lmr}^{iv} so as to minimize the total cost of shipments.[5] The objective function may be written as:

Objective function

$$\min Z(x,h) = \sum_i \sum_{lmrv} h_{lmr}^{iv} D_{lmr}^v p_{lmr}^{iv}. \tag{6.18}$$

Note that, unlike commodity flow models in the tradition of spatial-interaction modeling following from Wilson (1970) and discussed by Batten and Boyce (1986), we do *not* include an entropy expression, either as a condition the solution of the model must satisfy or as part of the objective function, because cross-hauling of commodities (as intermediate inputs) is explicitly modeled by Eq. (6.6).

The model can be solved as a mixed-complementarity problem. Noting that the input demand Eq. (6.15) already represents (has been derived from) a first-order condition for minimizing costs of production, and that the modified material balance Eq. (6.3′) and the gravity allocation of final demand Eq. (6.7) have been used to substitute into the modified conservation of flows constraint Eq. (6.4′), then using identity (6.2) to substitute for f_a^v, the relevant Lagrangian expression may be written as:

[4]Other characterizations of delivered pricing are offered in Chap. 7.

[5]Essentially, this is a problem of "traffic assignment" for carriers, given the demands for goods by firms and households at the nodes of the networks. But the influence of carriers' routing choices of commodity flows on delivered prices affects at least firms' demands for intermediate goods; so the problem may be viewed as a static cooperative game.

$$L(x,h) = \sum_i \sum_{lmrv} h^{iv}_{lmr} D^{v}_{lmr} p^{iv}_{lmr} + \sum_i \sum_{lmv} \beta^{iv}_{lm} \left(\sum_j \wp^{iv} x^{ij}_{lm} / q^i_x + \wp^{iv} \overline{\mathrm{FD}}^i_{lm} / q^i_x - \sum_r h^{iv}_{lmr} \right)$$
$$+ \sum_{ij} \sum_{lm} \phi^{ij}_{lm} \left(\frac{\varepsilon^{ij}_{lm} \kappa^j_m \alpha^{ij}_m}{\pi} \cdot \frac{P^j_m}{p^i_{lm\cdot}} \cdot X_m^{j(\rho^j_m+\kappa^j_m)/\kappa^j_m} c_m^{ij-\rho^j_m} - x^{ij}_{lm} \right)$$
$$+ \sum_{av} \psi^v_a \left(k^v_a - \sum_i \sum_{lmrv} h^{iv}_{lmr} \delta^{av}_{lmr} + TT^v_a \right), \tag{6.19}$$

where $c^{ij}_m, \wp^{iv}$, $p^{iv}_{lmr}, p^i_{lm\cdot}$, and $\overline{\mathrm{FD}}^i_{lm}$ are as defined above and β^{iv}_{lm}, ϕ^{ij}_{lm}, and ψ^v_a are the Lagrange multipliers associated with the modified conservation of flows constraints, the input demand equations, and the capacity flow constraints.

The Karush-Kuhn-Tucker conditions defining the saddle-point cost-minimizing solution are as follows:

$$\partial L / \partial h^{iv}_{lmr} = D^v_{lmr} p^{iv}_{lmr} - \beta^{iv}_{lm} + \phi^{ij}_{lm} x^{ij}_{lm} / p^i_{lm\cdot} \left[p^{iv}_{lmr} \left(\sum_{vr} h^{iv}_{lmr} - h^{iv}_{lmr} \right) / \left(\sum_{vr} h^{iv}_{lmr} \right)^2 \right]$$
$$- \psi^v_a \delta^{av}_{lmr} \geq 0, \quad \forall i, \forall l, \forall m, \forall r, \forall v. \tag{6.20}$$

$$\partial L / \partial x^{ij}_{lm} = \sum_v \left(\beta^{iv}_{lm} (\wp^{iv} / q^i_x) \left(1 - D^{*-1}_{lm} \sum_j X^j_m / \sum_m \sum_j X^j_m \right) \right)$$
$$- \phi^{ij}_{lm} \left(\frac{\rho^j_m \left(\varepsilon^{ij}_{lm}\right)^2 \kappa^j_m \alpha^{ij}_m}{\pi} \cdot \frac{P^j_m}{p^i_{lm\cdot}} \cdot X_m^{j(\rho^j_m+\kappa^j_m)/\kappa^j_m} c_m^{ij-\rho^j_m} / x^{ij}_{lm} + 1 \right) \geq 0, \forall i, \forall j, \forall l, \forall m. \tag{6.21}$$

$$h^{iv}_{lmr} \left(\partial L / \partial h^i_{lmrt} \right) = 0, \forall i, \forall v, \forall l, \forall m, \forall r, \tag{6.22}$$

$$x^{ij}_{lm} \left(\partial L / \partial x^{ij}_{lm} \right) = 0, \forall i, \forall j, \forall l, \forall m, \tag{6.23}$$

$$h^{iv}_{lmr} \geq 0, \forall i, \forall l, \forall m, \forall r, \forall v; x^{ij}_{lm} \geq 0, \forall i, \forall j, \forall l, \forall m. \tag{6.24}$$

$$\partial L / \partial \beta^{iv}_{lm} = \sum_j \wp^{iv} x^{ij}_{lm} / q^i_x + \wp^{iv} \overline{\mathrm{FD}}^i_{lm} / q^i_x - \sum_r h^{iv}_{lmr} \leq 0, \forall i, \forall j, \forall l, \forall m, \tag{6.25}$$

$$\partial L / \partial \phi^{ij}_{lm} = \frac{\varepsilon^{ij}_{lm} \kappa^j_m \alpha^{ij}_m}{\pi} \cdot \frac{P^j_m}{p^i_{lm\cdot}} \cdot X_m^{j(\rho^j_m+\kappa^j_m)/\kappa^j_m} c_m^{ij-\rho^j_m} - x^{ij}_{lm} \leq 0, \forall i, \forall j, \forall l, \forall m, \tag{6.26}$$

$$\partial L/\partial \psi_a^v = k_a^v - \sum_i \sum_{lmr} h_{lmr}^{iv} \delta_{lmr}^{av} + TT_a^v \leq 0, \forall a, \forall v, \tag{6.27}$$

$$\beta_{lm}^{iv}\left(\partial L/\partial \beta_{lm}^{iv}\right) = 0, \forall i, \forall j, \forall l, \forall m, \tag{6.28}$$

$$\phi_{lm}^{ij}\left(\partial L/\partial \phi_{lm}^{ij}\right) = 0, \forall i, \forall j, \forall l, \forall m, \tag{6.29}$$

$$\psi_a^v\left(\partial L/\partial \psi_a^v\right) = 0, \forall a, \forall v, \tag{6.30}$$

$$\beta_{lm}^{iv} \geq 0, \forall i, \forall l, \forall m, \forall v; \phi_{lm}^{ij} \geq 0, \forall i, \forall j, \forall l, \forall m; \psi_a^v \geq 0, \forall a, \forall v. \tag{6.31}$$

The model, which can be solved by the PATH algorithm of Dirkse and Ferris (1995), comprises conditions (6.20)–(6.31).

6.3 Extensions to a Dynamic Model of Location, Production, and Interaction

Batten and Boyce (1986) discuss under "new developments and prospects" for spatial interaction and commodity-flow models the integration of location, production, and interaction behavior (or LPI), noting that such modeling work is jointly concerned with location of physical capital stock, productive activities occurring with this stock, and flows generated by these activities. An example of LPI modeling is provided by Kohno et al. (1994)—henceforth KHM (1994)—who constructed a dynamic (multi-period) input–output programming model to measure socioeconomic effects of the (then) proposed Asian Expressway Network. The model not only explicitly incorporates industrial capital goods but also infrastructural facilities and incorporates the concept of interregional input–output programming of Moses (1960). Also embodied in the model is the computational technique of "simulation of traffic assignment on the network" (or STAN). The model can be employed to answer the question of what facilities should be invested in, when, where and by how much in order to maximize aggregate regional income. The authors employed the model to derive an investment criterion and conduct a feasibility study based on anticipated traffic volume.

The static commodity flow model developed in the previous section can be modified to embody some of the features of the KHM model—i.e., its dynamic formulation and endogenous determination of industrial capital stocks—as well as endogenous determination of labor employed and industrial output. In this latter formulation, the behavior of representative firms in each industry and location can be characterized by reaction functions, according to which firms adjust the capital stock, labor employed, intermediate input purchases, and output to their respective partial equilibrium levels. In the case of output, the partial equilibrium level is taken to be expected demand, $\widetilde{X}_m^j$.[6] These levels are—in most cases—themselves functions of prices set by firms (f.o.b. prices) or carriers (c.i.f. prices) or determined in other markets (e.g., markets for labor and transport services). The reaction functions are as follows:

$$\begin{aligned} \dot{x}_{lm}^{ij} &= \gamma_{lm}^{xij}\left(x_{lm}^{ij} * - x_{lm}^{ij}\right), \forall i, \forall j, \forall l, \forall m, \\ \text{where } x_{lm}^{ij} * &= \frac{\varepsilon_{lm}^{ij} \kappa_m^j \alpha_m^{ij}}{\pi} \cdot \frac{P_m^j}{p_{lm}^i} \cdot X_m^{j\left(\rho_m^j + \kappa_m^j\right)/\kappa_m^j} c_m^{ij-\rho_m^j}, \end{aligned} \tag{6.32}$$

$$\begin{aligned} \dot{L}_m^j &= \gamma_m^{Lj}\left(L_m^j * - L_m^j\right), \forall j, \forall m, \\ \text{where } L_m^j * &= \left[\frac{\kappa_m^j}{\pi} \cdot \alpha_m^{Lj} \cdot \frac{P_m^j}{w_m^j}\right]^{1/\left(1+\rho_m^j\right)} \left(X_m^j\right)^{\left(\kappa_m^j + \rho_m^j\right)/\left(\kappa_m^j + \kappa_m^j \rho_m^j\right)}, \end{aligned} \tag{6.33}$$

$$\begin{aligned} \dot{K}_m^j &= \gamma_m^{Kj}\left(K_m^j * - K_m^j\right), \forall j, \forall m, \\ \text{where } K_m^j * &= \left[\frac{\kappa_m^j}{\pi} \cdot \alpha_m^{Kj} \cdot \frac{P_m^j}{ucc_m^j}\right]^{1/\left(1+\rho_m^j\right)} \left(X_m^j\right)^{\left(\kappa_m^j + \rho_m^j\right)/\left(\kappa_m^j + \kappa_m^j \rho_m^j\right)}, \end{aligned} \tag{6.34}$$

$$\dot{X}_m^j = \gamma_m^{Xj}\left(\widetilde{X}_m^j - X_m^j\right). \tag{6.35}$$

The γ parameters appearing in Eqs. (6.32)–(6.35) are disequilibrium adjustment parameters. The partial-equilibrium level to which intermediate input demand, $x_{lm}^{ij}*$, adjusts is just the level derived in the static model above. The partial equilibrium levels to which the capital stock and labor employed adjust, $L_m^j *$ and $K_m^j *$, are derived in a manner similar to that of intermediate input demand. In the expressions for these quantities, w_m^j is the prevailing wage paid in industry j in location m, and ucc_m^j is the analogous user cost of capital. Expected sectoral output/demand is assumed to adjust to actual output/demand according to a process of adaptive expectations.

Given the reaction functions of representative firms (shippers), link capacities, and relevant prices, and the zero-order equations characterizing network flows in the static model, we can formulate a cooperative differential game between shippers and carriers in which the joint objective is to minimize the present value of shipping costs over a relevant planning horizon. The objective function of this game would be:

[6] We are assuming for the sake of convenience that forecasts of expected demand can be obtained (by survey or time-series forecasting) or projected by means of a forcing function of time. The formation of demand expectations can be endogenized, however, as in many macro-econometric studies, and the expectations variable eliminated by introducing a higher-order differential equation and substituting for this variable an expression written solely in terms of observable quantities and parameter estimates. See, e.g., Donaghy (1993) for a demonstration.

$$J = \min_{\{x,h\}} \int_{t_0}^{t_1} e^{-\lambda_c t} \sum_i \sum_{lmrv} h^{iv}_{lmr} D^{v}_{lmr} p^{iv}_{lmr} dt. \tag{6.36}$$

Following Kamien and Schwartz (1981), the intertemporal Lagrangian function for this optimization problem can be formulated as:

$$\begin{aligned}
L = {} & e^{-\lambda_c t} \sum_i \sum_{lmvr} h^{iv}_{lmr} D^{v}_{lmr} p^{iv}_{lmr} + \sum_{ij} \sum_{lm} \mu^{xij}_{lm} e^{-\lambda_c t} \left[\gamma^{xij}_{lm} \left(x^{ij}_{lm} * - x^{ij}_{lm} \right) - \dot{x}^{ij}_{lm} \right] \\
& + \sum_j \sum_m \mu^{Lj}_{m} e^{-\lambda_c t} \left[\gamma^{L_j}_{m} \left(L^{j}_{m} * - L^{j}_{m} \right) - \dot{L}^{j}_{m} \right] \\
& + \sum_j \sum_m \mu^{Kj}_{m} e^{-\lambda_c t} \left[\gamma^{K_j}_{m} \left(K^{j}_{m} * - K^{j}_{m} \right) - \dot{K}^{j}_{m} \right] \\
& + \sum_j \sum_m \mu^{\widetilde{X}j}_{m} e^{-\lambda_c t} \left[\gamma^{\widetilde{X}j}_{m} \left(X^{j}_{m} - \widetilde{X}^{j}_{m} \right) - \dot{\widetilde{X}}^{j}_{m} \right] \\
& + \sum_i \sum_{lmv} \zeta^{icon}_{lmv} e^{-\lambda_c t} \left(\sum_j \wp^{iv} x^{ij}_{lm} / q^{i}_{x} + \wp^{iv} \overline{FD}^{i}_{lm} / q^{i}_{x} - \sum_r h^{iv}_{lmr} \right) \\
& + \sum_a \sum_v \zeta^{flo}_{av} e^{-\lambda_c t} \left(\sum_i \sum_{lmr} h^{iv}_{lmr} \delta^{av}_{lmr} + TT^{v}_{a} - f^{v}_{a} \right) \\
& + \sum_a \sum_v \zeta^{cap}_{av} e^{-\lambda_c t} \left(k^{v}_{a} - f^{v}_{a} \right) + \sum_a \sum_v \zeta^{f}_{av} e^{-\lambda_c t} f^{v}_{a} \\
& + \sum_i \sum_{lmrv} \zeta^{hiv}_{lmr} e^{-\lambda_c t} h^{iv}_{lmr} + \sum_{ij} \sum_{lm} \zeta^{xij}_{lm} e^{-\lambda_c t} x^{ij}_{lm}.
\end{aligned} \tag{6.37}$$

In this expression, the μ variables are co-state (or adjoint) variables, whereas the ζ variables are intertemporal Lagrange multipliers.

In addition to initial values given for the state variables, the state Eqs. (6.32)–(6.35), and zero-order conservation of flows Eq. (6.4′), the first-order necessary conditions for an optimal solution to the dynamic optimization problem are as follows:

$$\begin{aligned}
\partial L / \partial x^{ij}_{lm} & = -\mu^{xij}_{lm} e^{-\lambda_c t} \gamma^{xij}_{lm} + \zeta^{icon}_{lmv} e^{-\lambda_c t} \left(\wp^{iv} / q^{i}_{x} \right) \left(1 + D^{*-1}_{lm} X^{j}_{m} \Big/ \sum_m \sum_j X^{j}_{m} \right) + \zeta^{xij}_{lm} \\
& = -\dot{\mu}^{xij}_{lm} e^{-\lambda_c t} + \lambda_c \mu^{xij}_{lm} e^{-\lambda_c t}, \forall i, \forall j, \forall l, \forall m,
\end{aligned} \tag{6.38}$$

$$\partial L / \partial L^{j}_{m} = -\mu^{Lj}_{m} e^{-\lambda_c t} \gamma^{Lj}_{m} = -\dot{\mu}^{Lj}_{m} e^{-\lambda_c t} + \lambda_c \mu^{Lj}_{m} e^{-\lambda_c t}, \forall j, \forall m, \tag{6.39}$$

$$\partial L/\partial K_m^j = -\mu_m^{Kj} e^{-\lambda_c t} \gamma_m^{Kj} = \dot{\mu}_m^{Kj} e^{-\lambda_c t} + \lambda_c \mu_m^{Kj} e^{-\lambda_c t}, \forall j, \forall m, \tag{6.40}$$

$$\begin{aligned} \partial L/\partial X_m^j &= \mu_{lm}^{xij} e^{-\lambda_c t} \gamma_{lm}^{xij} \left[\left(\rho_m^j + \kappa_m^j \right) / \kappa_m^j \right] x_{lm}^{ij} * / X_m^j \\ &+ \mu_m^{Lj} e^{-\lambda_c t} \gamma_m^{Lj} \left[\left(\kappa_m^j + \rho_m^j \right) / \left(\kappa_m^j + \kappa_m^j \rho_m^j \right) \right] L_m^j * / X_m^j \\ &+ \mu_m^{Kj} e^{-\lambda_c t} \gamma_m^{Kj} \left[\left(\kappa_m^j + \rho_m^j \right) / \left(\kappa_m^j + \kappa_m^j \rho_m^j \right) \right] K_m^j * / X_m^j \\ &+ \mu_m^{Xj} e^{-\lambda_c t} \gamma_m^{Xj} + \zeta_{lv}^{icon} e^{-\lambda_c t} \wp^{iv} \left(\partial \overline{FD}_{lm}^i / \partial X_m^j \right) / q_x^i \\ &= -\dot{\mu}_m^{Xj} e^{-\lambda_c t} + \lambda_c \mu_m^{Xj} e^{-\lambda_c t}, \forall \mathrm{j}, \forall \mathrm{m}, \end{aligned} \tag{6.41}$$

$$\text{where } \partial \overline{FD}_{lm}^i / \partial X_m^j = \frac{1}{\sum_m \sum_j X_m^j} \left\{ \left(X_l^i - \sum_m \sum_j X_m^j \right) D_{lmr}^{*-1} - \overline{FD}_{lm}^i \right\},$$

$$\begin{aligned} \partial L/\partial h_{lmr}^{iv} &= e^{-\lambda_c t} D_{lmr}^v p_{lmr}^{iv} + \mu_{lm}^{pi} e^{-\lambda_c t} \gamma_{lm}^{pi} \left[p_{lmr}^{iv} \left(\sum_{vr} h_{lmr}^{iv} - h_{lmr}^{iv} \right) / \left(\sum_{vr} h_{lmr}^{iv} \right)^2 \right] \\ &- \psi_{lmr}^{ij} \delta_{lmr}^{av} + \zeta_{lmr}^{hiv} = 0, \forall i, \forall j, \forall l, \forall m, \end{aligned} \tag{6.42}$$

$$\zeta_{av}^{cap} e^{-\lambda_c t} \left(k_a^v - f_a^v \right) = 0, \zeta_{av}^{cap} e^{-\lambda_c t} f_a^v = 0, \forall a, \forall v, \tag{6.43}$$

$$\zeta_{lmr}^{hiv} e^{-\lambda_c t} h_{lmr}^{iv} = 0, \forall i, \forall l, \forall m, \forall r, \forall v, \tag{6.44}$$

$$\zeta_{lm}^{xij} e^{-\lambda_c t} x_{lm}^{ij} = 0, \forall i, \forall j, \forall l, \forall m. \tag{6.45}$$

The transversality conditions that must be satisfied at every data point are as follows:

$$\begin{aligned} &\lim_{t \to \infty} \mu_{lm}^{xij} e^{-\lambda_c t} x_{lm}^{ij} = 0, \forall i, \forall j, \forall l, \forall m, \lim_{t \to \infty} \mu_m^{Lj} e^{-\lambda_c t} L_m^j = 0, \forall j, \forall m, \\ &\lim_{t \to \infty} \mu_m^{Kj} e^{-\lambda_c t} K_m^j = 0, \forall j, \forall m, \lim_{t \to \infty} \mu_m^{Xj} e^{-\lambda_c t} X_m^j = 0, \forall j, \forall m. \end{aligned} \tag{6.46}$$

The first-order necessary conditions can be solved by a variable step, variable order Adams method, as in the case of the dynamic network model of Donaghy and Schintler (1998). Discretized conditions can be solved by the PATH algorithm of Dirkse and Ferris (1995).

In this dynamic model, firms (shippers) react over time to prices and economies of scale and scope that can be realized by industrial location, thereby affecting the economic geography through their location decisions, while carriers react to shipping costs and firms' and households' demands for intermediate and final goods to adjust shipments between locations.[7]

[7]In Donaghy et al. (2005), changes in network (link) capacity are also considered in a dynamic optimization framework.

6.4 Conclusion

In this chapter, we have demonstrated how features of models in two distinct lineages of commodity-flow models—those of spatial interaction parentage and those of interregional input–output parentage—can be integrated into two new specifications that embody recent theoretical developments and extensions that permit examination of different classes of contemporary problems that are beyond the explanatory reach of previous models. In the next chapter, we discuss the econometric estimation of the parameters of a version of the dynamic models here presented with time-series data from 1977 to 2007 on commodity flows, whose derivation was discussed in Chap. 3.

References

Batten DF, Boyce D (1986) Spatial interaction, transportation, and interregional commodity flow models. In: Nijkamp P (ed) Handbook of regional and urban economics, vol I. North-Holland, Amsterdam, pp 357–406

Boyce D (2002) Combined model of interregional commodity flows on a transportation network. In: Hewings GJD, Sonis M, Boyce D (eds) Trade, networks, and hierarchies: modeling regional and interregional economies. Springer, Berlin, pp 29–40

Boyce D, Williams H (2015) Forecasting urban travel. Edward Elgar, Cheltenham

Dirkse SP, Ferris MC (1995) The PATH solver: a non-monotone stabilization scheme for mixed complementarity problems. Optim Methods Softw 5:123–156

Donaghy KP (1993) A continuous-time model of the United States economy. In: Gandolfo G (ed) Continuous time econometrics: theory and applications. Chapman and Hall, London, pp 151–194

Donaghy KD, Schintler LA (1998) Managing congestion, pollution, and pavement conditions in a dynamic transportation model. Transp Res D: Transp Environ 3(2):59–80

Donaghy K (2009) Modeling the economy as an evolving space of flows. In: Reggiani A, Nijkamp P (eds) Complexity and spatial networks. Springer, Berlin, pp 151–164

Donaghy K, Vial JF, Hewings GJD, Balta N (2005) A sketch and simulation of an integrated modeling framework for the study of interdependent infrastructure-based networked systems. In: Reggiani A, Schintler LA (eds) Methods and models in transport communications: cross Atlantic perspectives. Springer, Berlin, pp 93–117

Ham H, Kim TJ, Boyce D (2005) Implementation and estimation of a combined model of interregional, multimodal commodity shipments and transportation network flows. Transp Res B 39:65–79

Kamien MI, Schwartz NL (1981) Dynamic optimization: the calculus of variations and optimal control in economics and management. North Holland, New York

Kim TJ, Boyce DE, Hewings GJD (1983) Combined input-output and commodity flow models for interregional development planning: insights from a Korean application. Geogr Anal 15:330–342

Kohno H, Higano Y, Matsumura Y (1994) Dynamic takeoff-accelerating effects and the feasibility of the Asian expressway network on China. Paper presented at the 41st North American meetings of the RSAI, Niagara Falls, ON, Canada, November

Leontief WW, Strout A (1963) Multiregional input-output analysis. In: Barna T (ed) Structural interdependence and economic development. St. Martin's Press, New York, pp 119–150
Moses L (1960) A general equilibrium model of production, interregional trade, and location of industry. Rev Econ Stat 42:373–397
Southworth F, Peterson BE (2000) Intermodal and international freight network modeling. Transp Res C 8:147–166
Wilson AG (1970) Interregional commodity flows: entropy maximizing procedures. Geogr Anal 2:255–282

Chapter 7
Estimation of a Continuous-Time Structural-Equation Model of Commodity Flows

Abstract This chapter presents the operationalization and econometric estimation of a modified version of the dynamic continuous-time structural-equation model of commodity flows elaborated in the previous chapter. The objectives of this exercise are threefold: (1) to estimate an empirically based dynamic model that can accommodate the stylized facts of globalization noted earlier in this volume; (2) to determine whether or not a model that embodies a New Economic Geography formulation of production is supported by the data, and (3) develop and make available for other scholars regional economic data that supplement the commodity-flow data whose derivation and analysis have been discussed in Chaps. 3–5.

7.1 Introduction

In this chapter, we discuss the econometric estimation of a modified version of the continuous-time structural-equation model of commodity flows elaborated in the previous chapter. In engaging in this estimation exercise we have had several objectives. The first and foremost objective has been to estimate an empirically based dynamic model that slips the bonds of a static, constant-returns-to-scale fixed-coefficient specification so that it can accommodate, in general, the developments "on the ground" described in Chap. 2 and, in specific, the evolving stylized facts of the globalization experienced in the spatial economies of states in the Midwestern and Northeastern regions of the United States) over the three decades for which we have derived data (discussed in Chaps. 3 and 4). These stylized facts include increased fragmentation of production manifested in increasing intra-industry trade

---

The principal author and lead researcher for this chapter was Ziye Zhang.

Supplementary Information The online version of this chapter (https://doi.org/10.1007/978-3-030-78555-0_7) contains supplementary material, which is available to authorized users.

K. Donaghy et al., *The Co-evolution of Commodity Flows, Economic Geography, and Emissions*, Advances in Spatial Science,
https://doi.org/10.1007/978-3-030-78555-0_7

in intermediate inputs (including services) and the increasing transport intensity of all economic production and consumption.

A second objective has been to determine whether or not a dynamic commodity flow model based on a New Economic Geography (NEG) formulation of production that allows for increasing returns to scale (IRS) and imperfect competition fits the data well or finds empirical support and thereby suggests potential use of NEG-based models in exercises of forecasting and policy analysis.[1]

A third objective of this work has been to develop and make available regional economic data that supplement the commodity flow data (discussed in Chaps. 3–5) that other investigators may find useful for their own research or to augment going forward. Consequently, we dedicate a substantial amount of space in this chapter to discussing data sources, data transformations, and how we have dealt with missing data, discontinuous time series, and scaling issues that have arisen.

The model we estimate is one of the disequilibrium adjustments (measured in terms of percentage changes) in which natural logarithms of the levels of variables—capital stocks, employment, and intermediate inputs—adjust to the natural logarithms of "partial-equilibrium" levels of economic theory. The specifications of the partial-equilibrium levels derive from a two-stage production technology with increasing returns to scale, hence imperfect competition, which is compatible with NEG (e.g., Dixit-Stiglitz) formulations. In this model there are both upstream suppliers and downstream purchasers and "delivered prices" factor in demands for intermediate inputs.

In the econometric analysis we present in this chapter we employ a continuous-time methodology. We do so for the reasons that (1) there is no natural unit of temporal aggregation in macroeconomic phenomena; (2) we are dealing with a "mixed sample" of stocks, flows, end-of-period observations, and period averages, which continuous-time estimators can take into account through suitable data transformations, (3) in continuous-time estimation estimates of adjustment times are independent of the observational interval, and (4) estimates are "super-efficient" (Phillips 1991).[2]

[1]We do not argue that a model, such as the one we have estimated and the behavioral assumptions it embodies, can be "confirmed" as the "true" or best characterization of the evolution of commodity flows by econometric analyses, such as the one in which we engage. All models are abstractions that comprise assumptions made for the sake of argument and their logical implications. It is hoped that any model can provide analytical insights and support useful thought experiments. As logical constructions, though, models can neither be "true" nor "false"—only logically valid or invalid. The logical implications of some models are better supported by empirical data than others and some models, based on competing assumptions, may be "observationally equivalent." To assess a model, it must be compared to other models to determine which does a better job of supporting causal explanations of observed phenomena and which does not. Moreover, this determination is not strictly a matter of statistical testing but involves the history of argumentation within academic fields—of who needs to be persuaded of what, and what is deemed to be compelling evidence. (See Runde, 1998.)

[2]On these and other compelling reasons for using continuous-time estimation see Gandolfo (1981).

Table 7.1 State group

Region	Group	State(s)	BEA region	Centroid
1	RUS	Rest of the United States		Denver, CO
2	ILMIWI	Illinois + Michigan + Wisconsin	Great Lakes	Chicago, Il
3	INOH	Indiana + Ohio	Great Lakes	Columbus, OH
4	PA	Pennsylvania	Mideast	Harrisburg, PA
5	NJ	New Jersey	Mideast	New Brunswick, NJ
6	NY	New York	Mideast	Albany, NY
7	CT	Connecticut	New England	Hartford, CT
8	MARI	Massachusetts + Rhode Island	New England	Worcester, MA
9	MEVT	Maine + Vermont	New England	Portland, ME

For our purposes, we designate the BEA's Great Lakes region as being in the Midwestern United States and the BEA's Mideast and New England regions as constituting the Northeastern United States

Because the commodity flows (inter-industry, inter-, and intra-state sales of intermediate inputs) are so numerous, even in the case of the delimited number of regions we are considering, estimation of the model presents challenges to the implementation of maximum likelihood Gaussian estimation. We meet these challenges by adopting a "divide and conquer" approach in which we estimate the parameters of blocs of equations that are quasi-independent of each other in a high-performance computing environment. Specifically, the estimation of the large set of equations is accomplished efficiently by embedding a program that implements the nonlinear continuous-time FIML estimator of Wymer (1993), ESCONA, in an executive Python wrapper, which conducts the estimation of subsets of the model's equations via artificial intelligence algorithms.[3]

In Sect. 7.2, we discuss the data requirements of the model, and, in Sects. 7.3 and 7.4 discuss calibration and estimation of the model's equations and comment on results obtained. Section 7.5 presents overall conclusions.

7.2 Data Requirements

In this structural-equation estimation exercise, we study 13 industries active in 9 state groups over the period of 1977 to 2007. The 13 industries are as described in Chaps. 3 and 4. The 9 state groups reflect a geopolitical agglomeration scheme that is new to this chapter and is intended to make the estimation more manageable. Table 7.1 identifies the geopolitical groupings, Table 7.2 lists the sectoral scheme and presents the aggregation scheme for sectoral data. Table 7.2 also presents the concordance used to convert US BEA data published on the SIC (Standard Industrial

[3]Information about steering files and the executive Python wrapper are available from the authors upon request.

Table 7.2 Sectors in the Midwest-Northeast REIM

Sector	Sector title	SIC	NAICS
1	Agriculture, Forestry, and Fisheries	01, 02, 07, 08, 09	111–115
2	Mining	10, 12, 13, 14	21
3	Construction	15, 16, 17	23
4	Food and Kindred Products	20	311 + 312
5	Chemicals and Allied Products	28	325
6	Primary Metal Industries	33	331
7	Fabricated Metal Industries	34	332
8	Industrial Machinery and Equipment	35	333
9	Electronic and other Electric Equipment	36	334, 335
10	Transportation Equipment	37	336
11	Other Nondurable Manufacturing Products	21–23, 26, 27, 29–31	313–315; 322; 324; 511;516;324;326;316
12	Other Durable Manufacturing Products	24, 25, 32, 38, 39	321; 337; 327; 329
13	TCU, Service, and Government Enterprises	40–42, 44–65, 67, 70, 72, 73, 75, 76	42, 44, 45, 48, 49, 562, 51 (except 511, 516), 51–56; 61, 62, 71, 72, 81

Classification) and NAICS (North American Industry Classification System) sectors to consistent time series for the 13 sectors discussed in Chaps. 3, 4, and 5. Table 7.3 defines the variables which the commodity-flow model comprises.

To estimate the model, 1053 (9 state groups $\times$ 13 sector groups $\times$ 9 state groups) spreadsheets were prepared, each containing time-series data (31 annual observations, 1977–2007) for the variables in Table 7.3.

7.2.1 *Fixed Assets*

7.2.1.1 Data Source

The source of data on fixed assets is the US Bureau of Economic Analysis (hereafter, BEA) *Table 3.1ESI. Current-Cost Net Stock of Private Fixed Assets by Industry,*

Table 7.3 Variable descriptions

Variable	Description
K_m^j	Fixed assets in real-value terms of sector *j* at location *m*
L_m^j	Employment in sector *j* at location *m*
x_{lm}^{ij}	Intermediate inputs in real-value terms
X_m^j	Aggregate output in real-value terms of sector *j* at location *m*
c_m^{ij}	Input bundles used in production of sector *j* at location *m*
W_m^j	Average annual wage per employee in sector *j* at location *m* in nominal terms
P_m^j	GDP price deflator for sector *j* at location *m*
ucc_m^j	User cost of capital for sector *j* at location *m*
p_{lm}^i	Delivered prices corresponding to commodity flows x_{lm}^{ij}

which contains year-end estimates of the net stock of private assets[4] in billions of US dollars. These estimates were updated on August 8, 2019. According to the BEA,[5] "current-cost" means that an asset is valued at the price prevailing at the time the valuation is made. For example, the 2010 current-cost estimate for an asset is based on the price that would have been paid to acquire that asset in 2010. Before using the source data, we made several adjustments.

7.2.1.2 Data Deflation

Because the current-cost net stock data are in current values, to render them consistent with other variables, we deflated them to constant 2001 dollar values, using the GDP deflators discussed in Sect. 7.2.6. Each net stock is deflated as follows:

$$\text{NetStock}_{it}^{\text{deflated}} = \text{NetStock}_{it} \times \text{GDP deflator}_{it} \tag{7.1}$$

where $\text{NetStock}_{it}^{\text{deflated}}$ is the deflated net stock for industry *i* at time *t* in 2012 dollars; NetStock_{it} is the current-cost net stock for industry *i* at time *t*; GDP deflator_{it} is the GDP deflator (base year = 2001) for industry *i* at time *t*.

7.2.1.3 Allocation of the US Capital Stock to State Groups

Third, we need to allocate the national capital stock data to 9 state groups. Specifically, we split US net stock of private fixed capital for each sector *j* in year *t*, K_{USt}^j,

[4]For definition and details, see "Definitions and Introduction to Fixed Assets" from BEA https://www.bea.gov/resources/learning-center/definitions-and-introduction-fixed-assets

[5]https://www.bea.gov/help/glossary/current-cost

into the counterpart for each state group m, K^j_{mt}. We apply the following steps for the split:

- We employ time-series data on gross output time each sector j in each state group m, X^j_{mt}, to calculate each state group m's total output share in the United States for each sector j in each year t, i.e., X^j_{mt}/X^j_{USt}, where $X^j_{USt} = \sum_m X^j_{mt}$. (The derivation of the values of X^j_{mt} is discussed in Sect. 7.2.4.)
- For the starting year t_0, we split the national net stock into each state group using their total output share in that year.

$$K^j_{mt_0} = K^j_{USt_0} * X^j_{mt_0}/X^j_{USt_0} \tag{7.2}$$

- Then, for the following years, we modify the first-year base quantities by fractions of the national changes:

$$K^j_{mt_1} = K^j_{mt_0} + \left(K^j_{USt_1} - K^j_{USt_0}\right) * X^j_{mt_1}/X^j_{USt_1}$$

$$K^j_{mt_2} = K^j_{mt_1} + \left(K^j_{USt_2} - K^j_{USt_1}\right) * \frac{X^j_{mt_2}}{X^{j.}_{USt_2}}$$

and so on such that, we have the recursive relationship

$$K^j_{mt+1} = K^j_{mt} + \left(K^j_{USt+1} - K^j_{USt}\right) * \frac{X^j_{mt+1}}{X^j_{USt+1}}. \tag{7.3}$$

In this way, the sectoral capital stock of each state grouping changes by a proportion of the change in the national sectoral capital stock over time and will be consistent with the stylized facts of the BEA national data.

7.2.2 Employment

7.2.2.1 Data Source

The data source of employment includes (1) BEA *SA25 Total full-time and part-time employment by SIC industry* and (2) BEA *SA25N Total full-time and part-time employment by NAICS industry*. The former provides the estimates of employment for 1969–2001[6] by SIC industry, and the latter for 1998–2013 by NAICS industry.

[6]We only use time series from 1976 to 1999 for SIC-code data imputation.

Table 7.4 Employment data missing values

Type	SIC	NAICS	Total
(D)	37	184	221
(NA)	0	0	0
(L)	0	N/A	0
(T)	N/A	3	3
Total	37	187	224

The raw data are for the United States and all 13 relevant states (listed in Table 7.1) in disaggregated NAICS/SIC sectors.

7.2.2.2 Missing Data

Before aggregation, we conduct data imputation for missing values. There are four types of missing values in the raw data (see Table 7.4):

- (D) Not shown to avoid disclosure of confidential information, but the estimates for this item are included in the total.
- (NA) Data not available for this year.
- (L) Less than 10 jobs, but the estimates for this item are included in the total (SIC only).
- (T) Estimate for employment suppressed to cover corresponding estimate for earnings. Estimates for this item are included in the total (NAICS only).

We apply an "average-changing-rate" approach for imputation that involves three steps:

1. For each sector in each state, we calculate the rate of change between each two successive years. Note that the missing values will result in many missing rates of change.
2. For each sector in each year, we average the rate of change across all states. Such average rates of change can only be obtained for those sector-year cases without missing values.
3. We use the average rates of change to impute missing data. This requires that any missing data point (or period) must have at least one side adjacent to a non-missing value, such that the non-missing value could be used as a starting point to use the average change rate for imputation.

Before applying this approach, we ensure that there are no sectors in the raw data with missing values for all years, since there would be no non-missing value to start the imputation from. Our raw data, unfortunately, have several such cases. But fortunately, they all belong to one sector group "11 Other Non-durable Manufacturing Products." Thus, we can adjust and pre-apply our aggregation process to this sector group for all states to overcome this issue.

Consider the NAICS data as an example. As shown in Fig. 7.1, Vermont has two sectors with missing values for all years: "Leather and allied product manufacturing"

SA25N Total full-time and part-time employment by NAICS industry 1/

Bureau of Economic Analysis
State or DC

GeoFips	GeoName	LineCode	Description	1998	1999	2000	2001	2002	2003	2004	2005
50000	Vermont	530	Nondurable goods manufacturing	13872	14038	13984	13092	12519	11768	11616	11347
50000	Vermont	531	Food manufacturing	4165	4376	4519	4603	4565	4258	4310	4265
50000	Vermont	532	Beverage and tobacco product manufacturing	(T)	(T)	(T)	289	341	354	351	406
50000	Vermont	533	Textile mills	321	329	321	303	223	210	203	189
50000	Vermont	534	Textile product mills	(T)	(T)	(T)	294	271	252	164	150
50000	Vermont	535	Apparel manufacturing	1020	991	931	907	784	657	732	590
50000	Vermont	536	Leather and allied product manufacturing	(T)	(T)	(T)	(D)	(D)	(D)	(D)	(D)
50000	Vermont	537	Paper manufacturing	2008	1937	1914	1869	1631	1352	1338	1361
50000	Vermont	538	Printing and related support activities	2950	2834	2549	2275	2154	2061	2032	1878
50000	Vermont	539	Petroleum and coal products manufacturing	(T)	(T)	(T)	(D)	(D)	(D)	(D)	(D)
50000	Vermont	541	Chemical manufacturing	892	996	1032	1447	1362	1453	1417	1440
50000	Vermont	542	Plastics and rubber products manufacturing	1847	1882	1979	1070	1076	1036	942	942

LineCode	2006	2007	2008	2009	2010	2011	2012	2013	Sector Group
530	10942	11123	10841	10406	10260	10714	11848	12513	
531	4181	4489	4635	4665	4649	5065	5925	6413	4
532	437	479	404	387	433	461	563	632	11
533	150	151	119	125	123	131	123	159	11
534	147	146	118	(D)	103	121	112	111	11
535	563	589	504	547	517	447	566	499	11
536	(D)	(D)	(D)	(D)	(D)	(D)	(D)	(D)	11
537	1274	1224	1170	1125	1085	1052	716	715	11
538	1628	1560	1433	1384	1327	1221	1249	1246	11
539	(D)	(D)	(D)	(D)	(D)	(D)	(D)	(D)	11
541	1079	1022	1051	871	821	916	1224	1388	5
542	1362	1328	1252	1041	1054	1112	1199	1182	11

Fig. 7.1 Employment data with all-year missing values

SA25N Total full-time and part-time employment by NAICS industry 1/

Bureau of Economic Analysis
State or DC

GeoFips	GeoName	LineCode	Description	1998	1999	2000	2001	2002	2003	2004	2005
44000	Rhode Island	530	Nondurable goods manufacturing	25764	25256	24681	23134	21870	20880	(D)	19413

LineCode	2006	2007	2008	2009	2010	2011	2012	2013	Sector Group
530	19076	18486	17341	15625	15153	15184	14981	15200	

Fig. 7.2 NAICS employment data missing issue for Rhode Island

SA25 Total full-time and part-time employment by SIC industry 1/

Bureau of Economic Analysis
State or DC

GeoFips	GeoName	LineCode	Description	1976	1977	1978	1979	1980	1981	1982	1983	1984	1985	1986
23000	Maine	471	Chemicals and allied products	1127	1218	1411	1367	1400	1291	951	979	978	965	943

LineCode	1987	1988	1989	1990	1991	1992	1993	1994	1995	1996	1997
471	965	960	(D)	1013	1006	1188	1244	1327	1596	1725	1814

Fig. 7.3 SIC employment data missing issue for Maine

(Line Code 536) and "Petroleum and coal products manufacturing" (Line Code 539), both of which belong to sector group 11. Missing all-year values prevent us from using any imputation methods, and our following aggregation process—summing values from all disaggregated sectors that belongs to the same sector group—would result in underestimation. Therefore, for sector group 11, we adjust and pre-apply the aggregation process. We observe that the total value of "Nondurable goods manufacturing" (Line Code 530) is the sum of all group 11 sectors except for two: "Food manufacturing" (Line Code 531) for sector group 4 and "Chemical manufacturing" (Line Code 541) for sector group 5. In addition, these three rows have no missing values at all. Thus, we can simply calculate the sector group 11's aggregated values by subtracting the two exceptions from the total, which avoids using the rows with all-year missing values.

For consistency, we pre-apply this adjusted aggregation approach for all other states' sector group 11 for both NAICS and SIC data. Doing so requires that each state has no missing values in the three lines: "Nondurable goods manufacturing," "Food manufacturing," and "Chemical manufacturing." All states meet this requirement except for Rhode Island's NAICS data and Maine's SIC data. The Rhode Island's case is shown in Fig. 7.2. There is one missing observation in 2004 for "Nondurable goods manufacturing." We use a simple average of 2003 and 2005 values to impute the data point before applying the adjusted aggregation approach. The case of Maine is shown in Fig. 7.3. Similarly, we use the simple average of values in 1988 and 1990 to impute the missing point in 1989 in the "Chemicals and allied products" line.

After applying the adjusted aggregation process for sector group 11 to both NAICS and SIC data of all states, we are ready to implement the

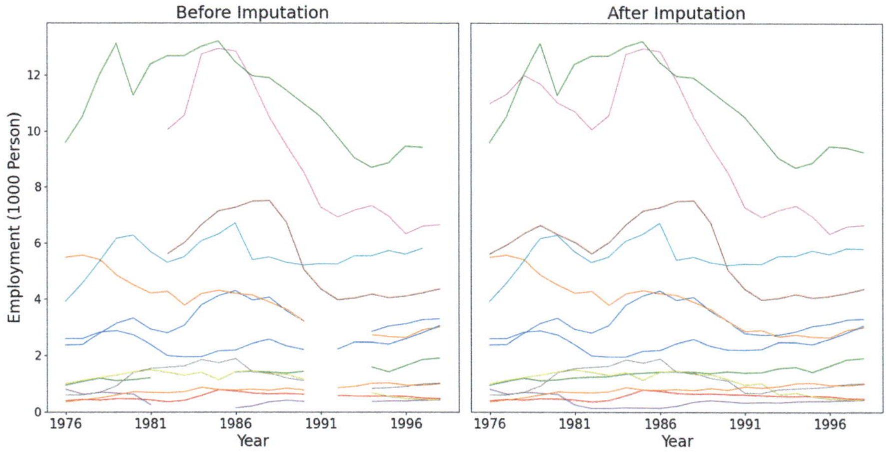

Fig. 7.4 Employment data imputation for SIC data

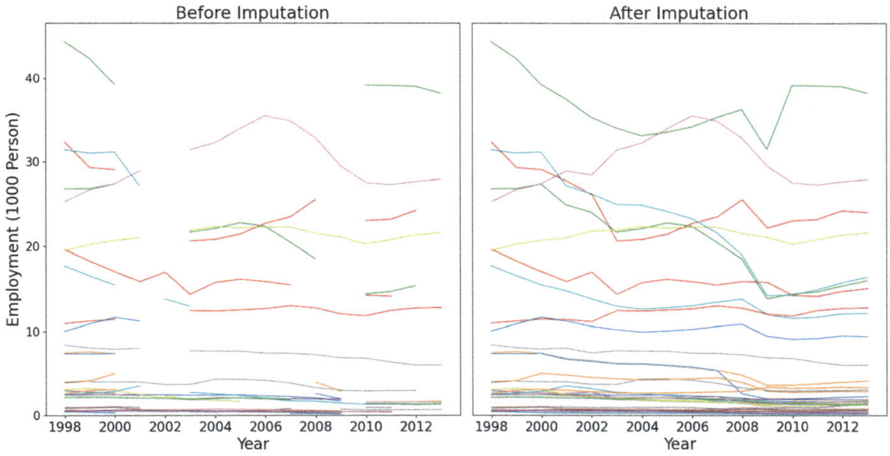

Fig. 7.5 Employment data imputation for NAICS data

"average-changing-rate" approach for data imputation. The number of missing observations of all relevant sectors are listed in Table 7.1. The comparison before and after the imputation is shown in Fig. 7.4 for SIC data and in Fig. 7.5 for NAICS data.

7.2.2.3 Data Aggregation over States and Sectors

After dealing with the missing values, we aggregate the data by state and sector groups according to Tables 7.1 and 7.2.

7.2.2.4 Splicing Time Series from SIC and NAICS Data Sets

The last step required is to splice time series from the SIC and NAICS data sets together. Although both data sets are aggregated into the same 13 sector groups and 9 state groups, the values are not comparable across these two different industrial classification periods. To combine the two data sets in a consistent way, we apply a backward extrapolation approach in the following steps:

1. For each sector-state, we calculate change rates between every two successive years during 1977–1998 using SIC data to capture the temporal variation of employment for 1977–1998.
2. Starting from 1998 NAICS data, we backward calculate new employment values year by year for 1977–1997 using the SIC change rates in the first step. In this way, we basically "shift" the SIC values to a comparable magnitude with NAICS without losing their original variations. See Fig. 7.6 for the comparison before and after adjustment for the Rest of the United States.

7.2.3 Intermediate Inputs

7.2.3.1 Data Source

Observations on inter-industry sales between 13 sector groups in 13 states and the rest of the United States were derived as discussed in Chap. 2. These data are separated into a set of MS Excel workbooks based on data year (e.g., A1.xlsx, A31.xlsx). Each workbook contains one master spreadsheet with 182 rows and 182 columns, reflecting all shipments among 13 sector groups in 13 states and the rest of the United States in that year. We aggregate the data into 9 state groups based on Table 7.1.

7.2.4 Gross Output

7.2.4.1 Data Source

Annual data on gross output in the United States of the 13 sectors modeled in millions of current dollars for 1977–2007 are taken from the US BEA database. These data are deflated by the US GDP price deflator for the base year of 2001, discussed in Sect. 7.5.

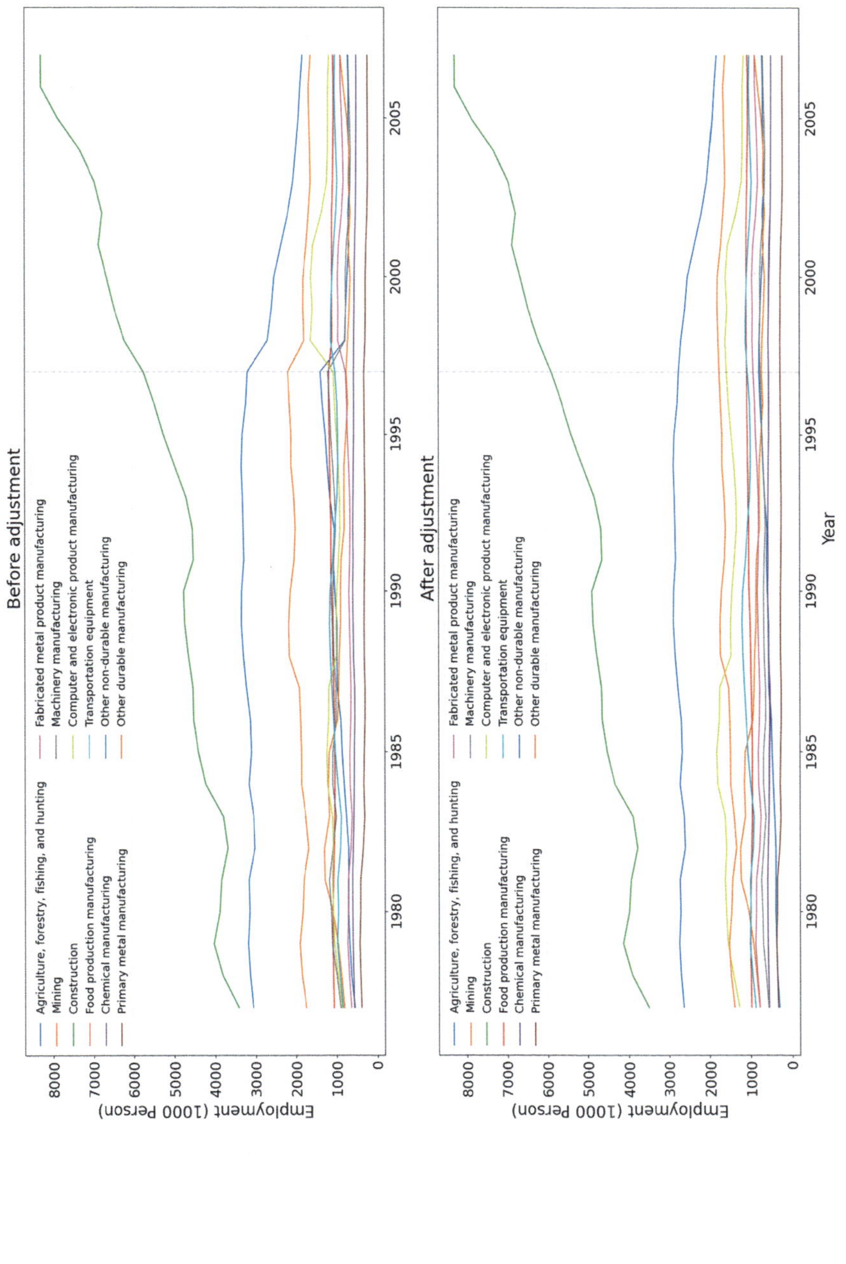

Fig. 7.6 Employment data SIC and NAICS adjustment example for rest of the United States

7.2.4.2 Data Split into State Groups

To produce the annual gross output data for each state group m in year t, X_{mt}^{j}, we multiply the gross output figures of sector j at the US national level in year t, X_{USt}^{j}, by the state group's share of total labor in the sector in the appropriate year, i.e.,

$$X_{mt}^{j} = X_{USt}^{j} \times \frac{L_{mt}^{j}}{\sum_{m} L_{mt}^{j}}, \tag{7.4}$$

where L_{mt}^{j} is the labor in sector j of state group m in year t; we use the employment data, discussed in Sect. 7.2.2, to measure labor.

7.2.5 *Wages*

7.2.5.1 Data Source

The sources of data on wages include (1) BEA *SA06 Compensation of employees by SIC industry* and (2) BEA *SA6N Compensation of Employees by NAICS Industry*. The former source provides the revised estimates of compensation for 1958–2001[7] by SIC industry, and the latter provides that for 1998–2013 by NAICS industry. The compensation in the raw data includes actual employer contributions and actuarially imputed employer contributions to reflect benefits accrued by defined benefit pension plan participants through service to employers in the current period. All dollar estimates are in current dollars (not adjusted for inflation). The raw data are for the United States and all 13 relevant states (listed in Table 7.1) in disaggregated NAICS/SIC sectors. The wage bill for each sector in each location is divided by the relevant number of employees to obtain an annual average wage figure. We then rescale the annual average wage figure to normalize it for estimation.

7.2.5.2 Missing Data

Similar to the employment data (discussed in Sect. 7.2.2), we conduct data imputation for missing values in wage data before aggregation. There are also four types of missing values:

- (D) Not shown to avoid disclosure of confidential information, but the estimates for this item are included in the total.
- (NA) Data not available for this year.

[7]We only use SIC-code time series from 1976 to 1999 for data imputation.

SA25N Total full-time and part-time employment by NAICS industry 1/

Bureau of Economic Analysis
State or DC

GeoFips	GeoName	LineCode	Description	1998	1999	2000	2001	2002	2003	2004	2005
50000	Vermont	530	Nondurable goods manufacturing	13872	14038	13984	13092	12519	11768	11616	11347
50000	Vermont	531	Food manufacturing	4165	4376	4519	4503	4555	4258	4310	4265
50000	Vermont	532	Beverage and tobacco product manufacturing	(T)	(T)	(T)	289	341	354	351	405
50000	Vermont	533	Textile mills	321	329	321	303	223	210	203	189
50000	Vermont	534	Textile product mills	(T)	(T)	(T)	294	271	252	164	150
50000	Vermont	535	Apparel manufacturing	1020	991	931	907	784	657	732	590
50000	Vermont	536	Leather and allied product manufacturing	(T)	(T)	(T)	(D)	(D)	(D)	(D)	(D)
50000	Vermont	537	Paper manufacturing	2008	1937	1914	1869	1631	1352	1338	1361
50000	Vermont	538	Printing and related support activities	2950	2834	2549	2275	2154	2061	2032	1878
50000	Vermont	539	Petroleum and coal products manufacturing	(T)	(T)	(T)	(D)	(D)	(D)	(D)	(D)
50000	Vermont	541	Chemical manufacturing	892	996	1032	1447	1362	1453	1417	1440
50000	Vermont	542	Plastics and rubber products manufacturing	1847	1882	1979	1070	1076	1036	942	942

LineCode	2006	2007	2008	2009	2010	2011	2012	2013	Sector Group
530	10942	11123	10841	10406	10260	10714	11848	12513	
531	4181	4489	4635	4665	4649	5065	5925	6413	4
532	437	479	404	387	433	461	563	632	11
533	150	151	119	125	123	131	123	159	11
534	147	146	118	(D)	103	121	112	111	11
535	563	589	504	547	517	447	566	499	11
536	(D)	(D)	(D)	(D)	(D)	(D)	(D)	(D)	11
537	1274	1224	1170	1125	1085	1052	716	715	11
538	1628	1550	1433	1384	1327	1221	1249	1246	11
539	(D)	(D)	(D)	(D)	(D)	(D)	(D)	(D)	11
541	1079	1022	1051	871	821	916	1224	1388	5
542	1352	1328	1252	1041	1054	1112	1199	1182	11

Fig. 7.7 Wage data with all-year missing values

SA06N Compensation of employees by NAICS industry

Bureau of Economic Analysis
State or DC

GeoFips	GeoName	LineCode	Description	1998	1999	2000	2001	2002	2003	2004	2005
44000	Rhode Island	530	Nondurable goods manufacturing	1001708	1026045	1033832	1001585	1005930	1019790	(D)	1016598

LineCode	2006	2007	2008	2009	2010	2011	2012	2013	Sector Group
530	1038706	1067984	1011632	897164	919381	926590	932685	964818	

Fig. 7.8 NAICS wage data missing issue for Rhode Island

SA06 Compensation of employees by SIC industry

Bureau of Economic Analysis
State or DC

GeoFips	GeoName	LineCode	Description	1976	1977	1978	1979	1980	1981	1982	1983	1984	1985	1986
09000	Connecticut	463	Food and kindred products	(D)	(D)	(D)	(D)	(D)	(D)	(D)	(D)	(D)	(D)	(D)
23000	Maine	471	Chemicals and allied products	14488	16784	20748	22915	25082	26676	21285	23460	25152	25003	25519
25000	Massachusetts	463	Food and kindred products	(D)	(D)	(D)	(D)	(D)	(D)	(D)	(D)	(D)	(D)	(D)

LineCode	1987	1988	1989	1990	1991	1992	1993	1994	1995	1996	1997
463	(D)	(D)	411844	434074	475050	456774	447378	463093	463698	421877	425840
471	(D)	(D)	(D)	(D)	(D)	48291	53391	56267	70118	80106	85639
463	(D)	(D)	663031	709643	717302	750920	781260	797866	814872	879363	913627

Fig. 7.9 SIC wage data missing issue for three states

- (L) Less than $50,000, but the estimates for this item are included in the total (SIC only).
- (T) Estimate for compensation suppressed to cover corresponding estimate for earnings. Estimates for this item are included in the total (NAICS only).

What we apply for imputation is the same "average-changing-rate" approach as used in the case of employment data. As mentioned before, this method does not work for state-sectors with missing values for all years. We have such cases in both NAICS and SIC wage data. Fortunately, they all belong to the sector group "11 Other Non-durable Manufacturing Products" as well. Thus, we follow the employment data procedure to adjust and pre-apply aggregation process to the wage data.

Consider the NAICS data as an example. As shown in Fig. 7.7, Vermont has two sectors with missing values for all years: "Leather and allied product manufacturing" (LineCode 536) and "Petroleum and coal products manufacturing" (LineCode 539), both of whom belong to sector group 11. Similarly, we calculate the sector group 11's aggregated values by subtraction and thus avoid using the rows with all-year missing values.

While applying this approach to all other states' sector group 11, we encounter several problematic cases in which values are missing in either of the three lines used in the subtraction. In the NAICS data, the case is Rhode Island. As shown in Fig. 7.8, since only the year 2004 is missing for this time series, we use a simple average of the values in 2003 and 2005 to impute the data and then apply the adjusted aggregation process for this state. In the SIC data, Connecticut, Maine, and Massachusetts have missing values in either "Food and kindred products" or "Chemicals and allied products" lines (see Fig. 7.9). In each case multiple-year values are

Table 7.5 Wage data missing values

Type	SIC	NAICS	Total
(D)	118	184	202
(NA)	0	0	0
(L)	0	N/A	0
(T)	N/A	3	3
Total	118	187	205

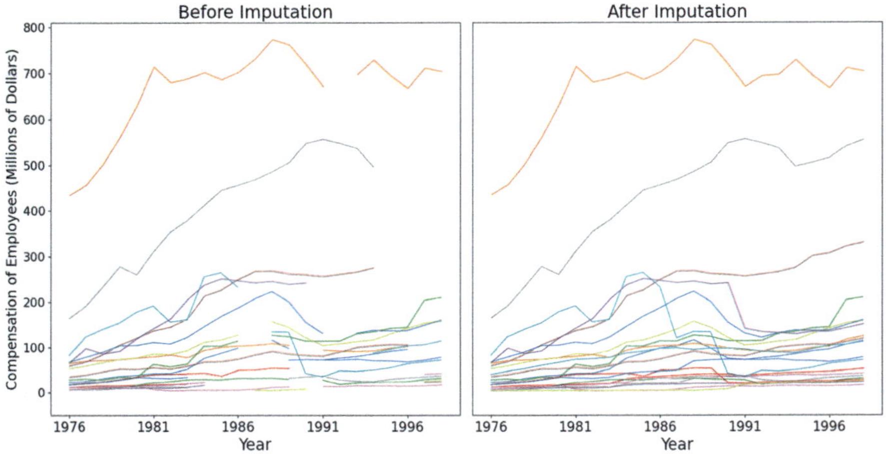

Fig. 7.10 Wage data imputation for SIC data

missing; hence we impute the values of these data using the "average-changing-rate" approach (introduced in Sect. 7.2.2.2) before the subtraction.

After fixing the problematic cases and applying adjusted aggregation process for sector group 11 of all states, we impute missing values in our raw data using the "average-changing-rate" approach. The number of missing observations of all relevant sectors are listed in Table 7.5. The comparison before and after the imputation is shown in Fig. 7.4 for SIC data and in Fig. 7.5 for NAICS data (Figs. 7.10 and 7.11).

7.2.5.3 Data Aggregation over States and Sectors

After dealing with the missing values, we aggregate the data to state and sector groups according to Tables 7.1 and 7.2.

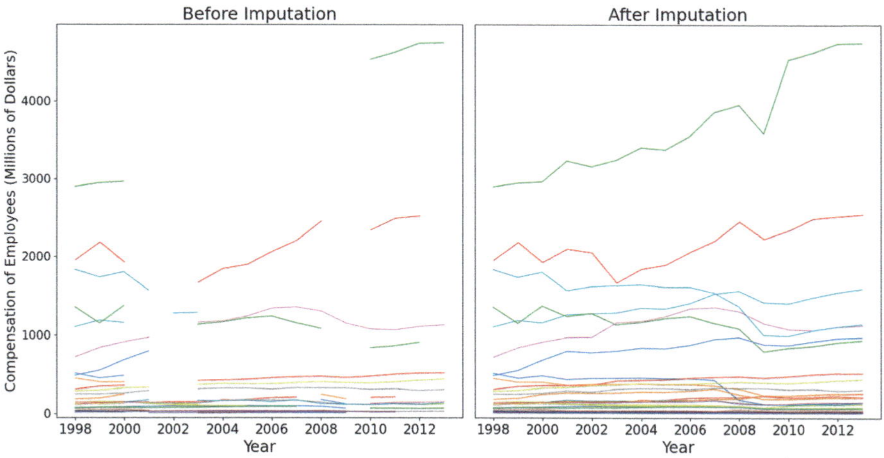

Fig. 7.11 Wage data imputation for NAICS data

7.2.5.4 Splicing SIC and NAICS Data

We follow the same backward extrapolation approach discussed in Sect. 7.2.2.4 to splice SIC and NAICS compensation data. See Fig. 7.12 for the comparison before and after adjustment for the Rest of US state groups.

7.2.6 Output Price

7.2.6.1 Data Source

For output price, P_m^j, we use the Gross Domestic Product (hereafter, GDP) deflator (base year 2001) for each sector j at each location m as a proxy for 1977 through 2007. The GDP deflator is the ratio of GDP in current dollars to GDP in 2001 dollars. The data sources of nominal GDP include (1) BEA *SAGDP2S Gross domestic product (GDP) by state* and (2) *SAGDP2N Gross domestic product (GDP) by state*. The former provides GDP in millions of current dollars for 1963–1997[8] by SIC industry, while the latter provides that for years after 1997 by NAICS industry.

[8]We only use time series from 1976 to 1997.

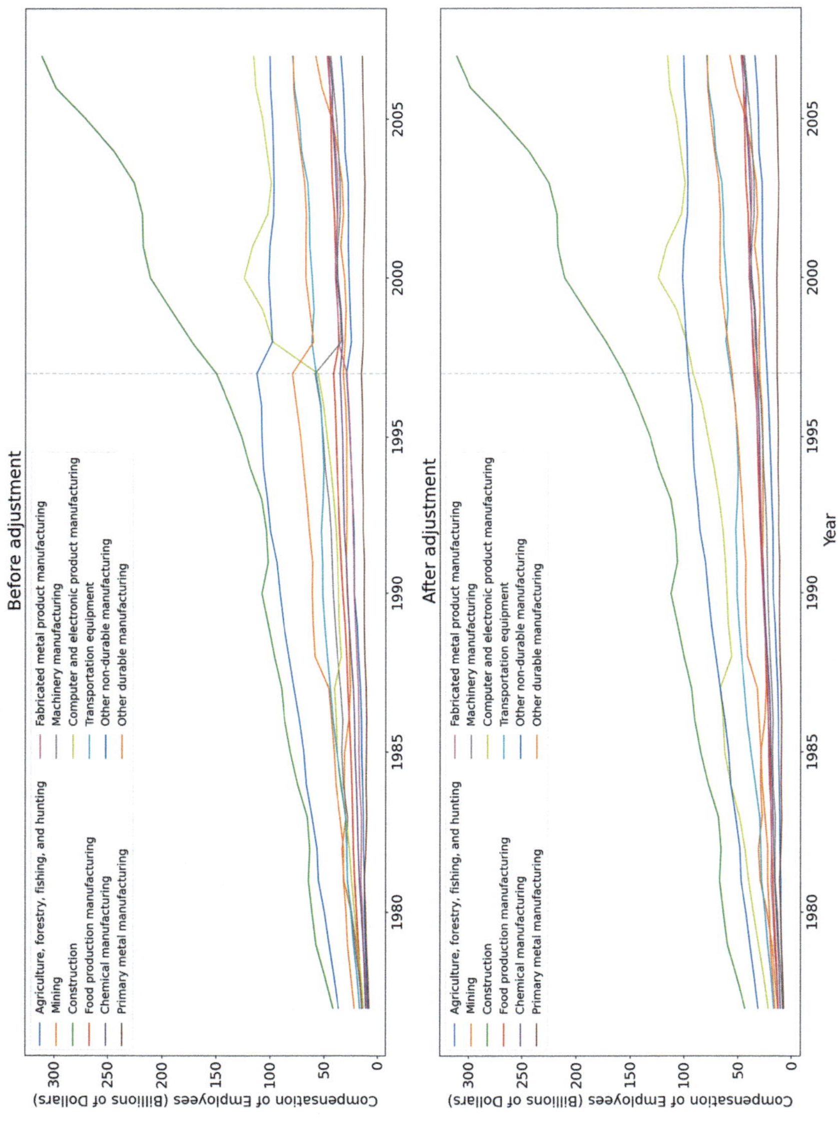

Fig. 7.12 Wage data SIC and NAICS adjustment example for rest of the United States

7.2.6.2 Data Aggregation over States and Sectors

First, we aggregate 13 states and the rest of the United States into 9 state groups based on Table 7.1 and merge NAICS/SIC sectors into 13 sector groups based on Table 7.2.

7.2.6.3 Data Splicing

Second, we splice time series on deflators in SIC years (1977–1997) with those in NAICS years (1997–2007). Note that since the SIC and NAICS raw data use different base years (SIC based on 1997 but NAICS on 2012), we convert NAICS base year to 1997 before the combination.

7.2.6.4 Base Year

To be consistent with intermediate inputs data, we further change the base year of the GDP deflators from 1997 to 2001. The GDP deflators for all sector groups of all state groups from 1977 to 2007 are shown in Fig. 7.13.

7.2.7 *User Cost of Capital*

The user cost of capital ucc_{mt}^{j} for sector j at location m at time t is calculated as:

$$\mathrm{ucc}_{mt}^{j} = PPI \times \left(\frac{\mathrm{InterestRate}_t}{100} + \mathrm{DepreciationRate}_t - \mathrm{InflationRate}_t\right), \tag{7.5}$$

where

- PPI is the producer price index of capital equipment from the US Bureau of Labor Statistics *Table Producer Price Index: Finished Goods: Capital Equipment*. The original index is based on 1982. We convert the base year to 2001 to be consistent with other data.
- Interest Rate is from Board of Governors of the Federal Reserve System *Table Market yield on US Treasury securities at 30-year constant maturity, quoted on investment basis*.
- Depreciation Rate is the ratio of capital depreciation to total capital stock. The capital depreciation is from BEA *Table 3.4ESI. Current-Cost Depreciation of Private Fixed Assets by Industry* in billions of dollars; while the total capital stock is from *BEA Table 3.1ESI. Current-Cost Net Stock of Private Fixed Assets by Industry* in billions of dollars.

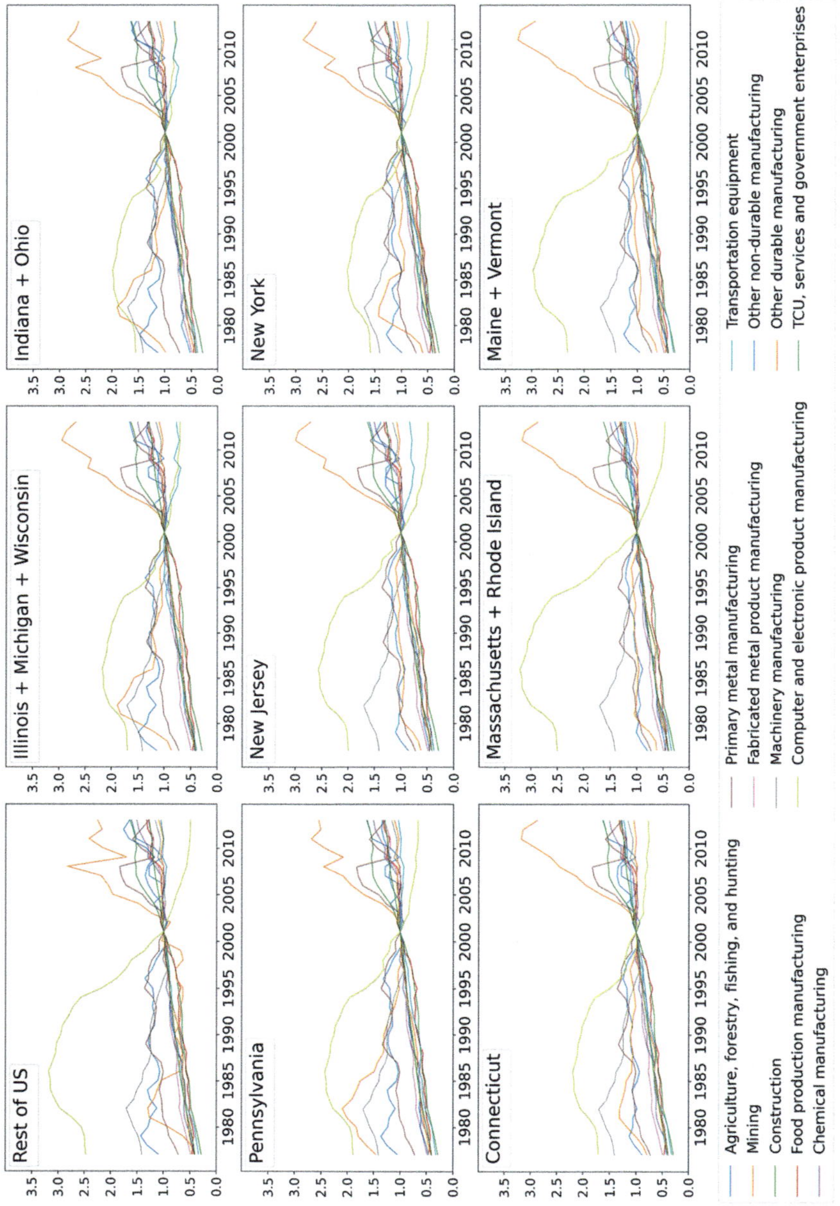

Fig. 7.13 GDP Deflator 1977–2007 (2001=100)

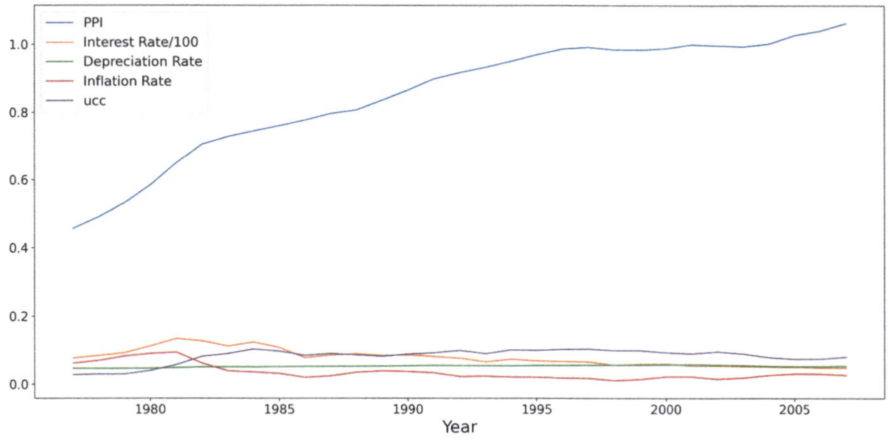

Fig. 7.14 User cost of capital data

- Inflation Rate is calculated using GDP price index numbers from BEA National Income and Product Accounts *Table 1.1.4. Price Indexes for Gross Domestic Product*. The inflation rate for each year is simply the percent change in price index from preceding year.

Note that since all data sources only vary by year, we assume ucc^{j}_{mt} does not change across states or sectors. The time series of all relevant data are tabulated in Appendix 7.1 and displayed in Fig. 7.14.

7.2.8 Delivered Price

7.2.8.1 Calculation

The delivered price at location m of sector good i produced at location l, p^{i}_{lmt} (also known as the price comprising the costs of carriage, insurance, and freight (c.i.f)), equals the price of the commodity at the factory gate P^{i}_{lt} (i.e., the *free-on-board* (f.o. b.) price or the output price discussed in Sect. 7.2.6, plus an integrated unit cost of transport good i by two transportation modes—truck and rail—from location l to location m, ϑ^{i}_{lmt},

$$p^{i}_{lmt} = P^{i}_{lt} + \vartheta^{i}_{lmt}, \forall l, \forall m, \forall i. \tag{7.6}$$

The integrated unit cost ϑ^{i}_{lm} is calculated as:

$$\vartheta_{lmt}^{i} = d_{lm}\sum_{r}\left(c_{t}^{r}\omega_{lmt}^{ir}\right)/q_{x}^{i}, \forall l, \forall m, \forall i. \tag{7.7}$$

where

- d_{lm} is the shortest distance between l and m. Distances among state groups are shown in Table 7.6. They are calculated based on centroid points in Table 7.1.
- c_{t}^{r} is the shipping cost per ton-mile by mode r in year t. The shipping cost is represented by transportation fee. The original data is from the US Bureau of Transportation Statistics *Table 3-21: Average Freight Revenue Per Ton-mile (current cents)*. However, missing data problem exists for both Truck and Rail data.
- Figure 7.15 displays the raw data table. For the study period 1977–2007, the truck data miss values for all years before 1990; while the rail data are only slightly better with two data points in 1980 and 1985. We have to impute data for the missing periods.
- For the rail data (see Fig. 7.16), since we have two anchor points within the missing period (1977–1989) and the whole series seems to follow a smooth trend, so we simply apply two linear growth models for imputation: one for 1985–1990, and the other for years prior to 1985. For the truck data, things look much more complicated. The entire missing period has no reference points, and the non-missing period seems to be volatile. Thus, we use a moving average approach to take both seasonality and trend into consideration. After testing, with moving window size being nine, the prediction errors are the smallest. Figure 7.16 shows a comparison between the raw data and the imputed data.
- ω_{lmt}^{ir} is the weighting factor of mode r of sector i from l to m in year t and calculated from a logit model,

$$\omega_{lmt}^{ir} = \frac{exp\left(b_{r}^{x}x_{lmt}^{ir} + b_{r}^{d}d_{lm} + \delta^{r}D^{i}\right)}{\sum_{r} exp\left(b_{r}^{x}x_{lmt}^{ir} + b_{r}^{d}d_{lm} + \delta^{r}D^{i}\right)}, \tag{7.8}$$

where x_{lmt}^{ir} is the values of shipments by mode r, and b_{r}^{x} and b_{r}^{d} are mode choice coefficients, δ^{r} is a dummy variable equal to 1 if the mode is truck, D^{i} is mode choice constant for sector i. We use the mode choice coefficients and constants provided by Ham et al. (2005) as shown in Table 7.7.

However, we have no data for shipments by different modes x_{lmt}^{ir}. Instead, we use value-to-weight conversion factors q_{x}^{i}, which only vary across sector groups and are constant across state groups and years. Thus, Eq. (7.8) becomes

$$\omega_{lm}^{ir} = \frac{exp\left(b_{r}^{x}q_{x}^{i} + b_{r}^{d}d_{lm} + \delta^{r}D^{i}\right)}{\sum_{r} exp\left(b_{r}^{x}q_{x}^{i} + b_{r}^{d}d_{lm} + \delta^{r}D^{i}\right)}. \tag{7.9}$$

Table 7.6 Distance table (in kilometers)

	RUS	ILMIWI	INOH	PA	NJ	NY	CT	MARI	MEVT
RUS	2000.0	2818.9	3432.5	3899.7	4126.5	4148.3	4239.9	4492.2	4717.8
ILMIWI	2818.9	300.0	519.4	1236.5	1228.3	1250.1	1341.7	1594.0	1819.6
INOH	3432.5	519.4	200.0	717.1	708.9	730.7	822.3	1074.6	1300.2
PA	3899.7	1236.5	717.1	300.0	226.8	248.6	340.2	592.5	818.1
NJ	4126.5	1228.3	708.9	226.8	100.0	21.8	113.4	365.7	591.3
NY	4148.3	1250.1	730.7	248.6	21.8	150.0	91.6	343.9	569.5
CT	4239.9	1341.7	822.3	340.2	113.4	91.6	50.0	252.3	477.9
MARI	4492.2	1594.0	1074.6	592.5	365.7	343.9	252.3	75.0	236.2
MEVT	4717.8	1819.6	1300.2	818.1	591.3	569.5	477.9	236.2	200.0

Table 3-21: Average Freight Revenue Per Ton-mile (Current ¢)

	1960	1965	1970	1975	1980	1985	1990	1991	1992	1993	1994	1995
Truck[b]	U	U	U	U	U	U	12.88	12.46	12.96	13.12	13.56	13.5
Class I rail	1.40	1.27	1.43	2.04	2.87	3.04	2.66	2.59	2.58	2.52	2.49	2.40

	1996	1997	1998	1999	2000	2001	2002	2003	2004	2005	2006	2007
Truck[b]	13.98	14.27	12.89	13.14	13.75	13.31	13.09	13.33	14.24	15.53	16.52	16.54
Class I rail	2.35	2.40	2.34	2.28	2.26	2.24	2.26	2.28	2.35	2.62	2.84	2.99

KEY: R = revised; U = data are unavailable.

Fig. 7.15 Average freight revenue raw data

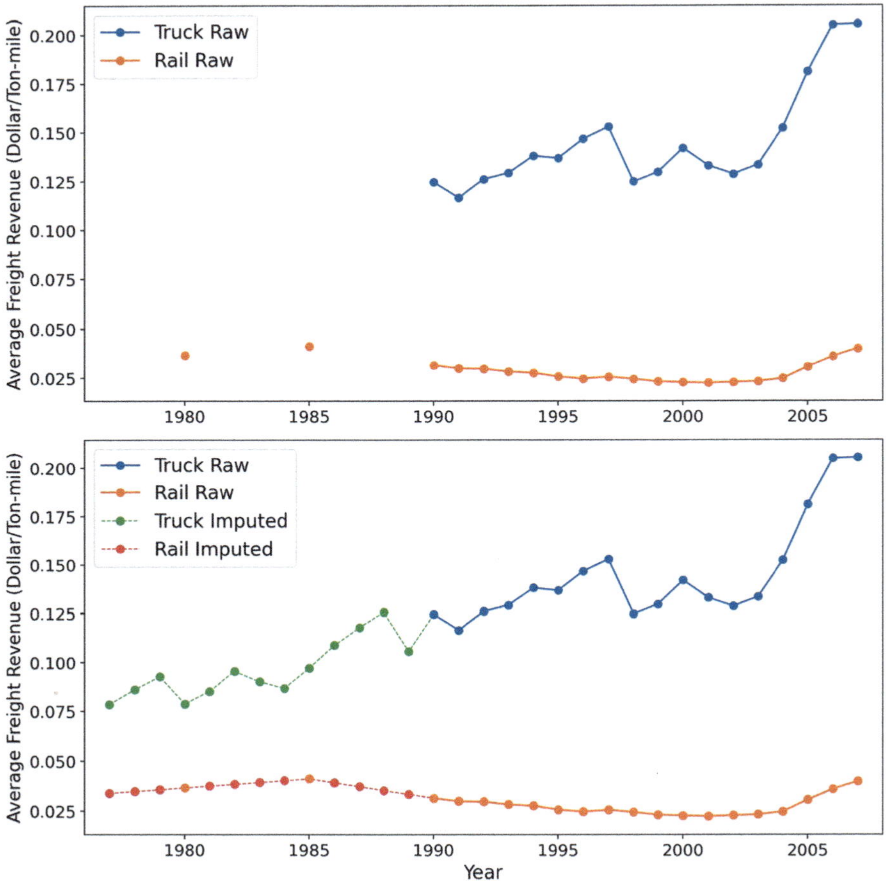

Fig. 7.16 Average freight revenue data imputation

- q_x^i is a factor converting millions of constant year-2001 dollars of output to kilotons.[9]

[9]Conversion factors in millions of constant year-2001 US dollars per kiloton for the 13 sectors are respectively 0.888524, 0.262133, 0.181015, 1.403953, 2.012055, 1.180493, 0.061437, 9.415343, 22.01849, 7.725906, 1.560004, 1.631441, and 3.103150 (USDOT Commodity Flow Survey for 2007).

Table 7.7 Mode choice coefficients

Coefficient	Estimated value
b^x_{truck}	+0.375
b^d_{truck}	−2.457
b^x_{rail}	−0.0754
b^d_{rail}	+0.29
D_1	1.131
D_2	1.067
D_3	0
D_4	3.523
D_5	2.503
D_6	3.043
D_7	4.038
D_8	1.567
D_9	1.661
D_{10}	0.659
D_{11}	2.612
D_{12}	3.095
D_{13}	0

7.2.8.2 Scale Issues

In calculating the delivered prices, we encounter two scale issues:

- First, in Eq. (7.9), the conversion factor q^i_x and distance d_{lm} differ in scale, we have to re-scale them before combination. To generate reasonable weighing factors, we experiment combinations of scaling schemes and find out that the best scaling choice is to divide the conversion factor by 10 ($q^i_x/10$) and to divide the distance by 10,000 (d_{lm}/10,000). This choice is the best since it produces share estimates closest to those observed in Ham et al. (2005). See Appendix 7.1 for more details of this experiment.
- Second, in Eq. (7.6), the GDP deflators P^i_l and the unit costs ϑ^i_{lm} are not comparable but we need to make sure that they follow some stylized facts of the transportation cost share of delivered costs. After consulting the costs on shipping coal (EIA 2020), which is among the heaviest of commodities shipped, we impose a ceiling on transportation costs of no more than 50% of the output price, that is,

$$p^i_{lmt} = P^i_{lt} + \epsilon^i_{lmt}\vartheta^i_{lmt}, \forall l, \forall m, \forall i, \tag{7.10}$$

where, $\epsilon^i_{lmt} = \begin{cases} 1 & \text{if } \vartheta^i_{lmt} \leq 0.5P^i_{lt} \\ 0.5P^i_{lt}/\vartheta^i_{lmt} & \text{if } \vartheta^i_{lmt} > 0.5P^i_{lt} \end{cases}$ where ϵ^i_{lmt} is a re-scaling factor to convert the unit costs ϑ^i_{lm} whose share in the output price is greater than 0.5 back to 0.5. We can see from Table 7.8 that most of the unit costs are originally in reasonable scale. Only about 5% of the unit costs got capped to one-half share.

Table 7.8 Distribution of unit cost shares in output prices

	Output price	Unit cost	Share
	P^i_{lt}	ϑ^i_{lmt}	$\vartheta^i_{lmt}/P^i_{lt}$
Mean	0.974	0.095	11.00%
Std	0.415	0.139	15.00%
Min	0.281	0.000	0.00%
25%	0.698	0.008	0.90%
50%	0.957	0.029	3.50%
70%	1.085	0.125	14.30%
80%	1.156	0.172	20.20%
85%	1.244	0.228	26.70%
90%	1.396	0.307	43.50%
95%	1.774	0.433	50.00%
96%	1.914	0.446	50.00%
97%	2.062	0.487	50.00%
98%	2.258	0.500	50.00%
99%	2.620	0.531	50.00%
Max	3.206	1.115	50.00%

Since conversion factor and distance differ in scale, to generate reasonable weighing factors, we experiment with combinations of scaling schemes on q^i_x and d_{lm} and find that the best scaling choice is that the conversion factor divided by 10 ($q^i_x/10$) and the distance divided by 10,000 ($d_{lm}/10{,}000$), which produces share estimates closest to the observed ones in Ham et al. (2005). See Appendix 7.1 for more details of this experiment.

7.3 Calibration of a Cobb-Douglas Aggregator Function for Intermediate Goods

As noted in Chap. 6, the production technology we are assuming to characterize industrial production in each sector in each location is a two-stage C.E.S. function in which intermediate inputs are aggregated at the first (or lower) stage by Cobb-Douglas functions, c^{ij}_{mt}. That is,

$$X^j_m = \left[\sum_i \alpha^{ij}_m c^{ij}_m{}^{-\rho^j_m} + \alpha^{Lj}_m \left(L^j_m e^{\lambda^j_m \cdot t}\right)^{-\rho^j_m} + \alpha^{Kj}_m K^j_m{}^{-\rho^j_m}\right]^{-\frac{\kappa^j_m}{\rho^j_m}}. \tag{7.11}$$

We begin the process of estimating the parameters of the continuous-time structural-equation commodity-flow model, by calibrating the share parameters of the Cobb-Douglas aggregator function for intermediate inputs at the first stage of the two-stage production function. This aggregator function is:

$$c_{mt}^{ij} = \gamma_m^{ij} \prod_l x_{lmt}^{ij\ \varepsilon_{lm}^{ij}}, \tag{7.12}$$

where c_{mt}^{ij} denotes the aggregate of inputs i shipped from all other locations to sector j at location m at time t, x_{lmt}^{ij} denotes inter-industry sales from sector i at location l to sector j and location m at time t, γ_m^{ij} is the scale parameter for the aggregator of inputs i in the production of good j at location m, and ε_{lm}^{ij} is the share parameter for input x_{lmt}^{ij}. Values of the scale parameter are not required for the estimation work discussed below but values of the share parameters are.

7.3.1 Data Requirements

To calibrate the share parameters in Eq. (7.12), we need data on intermediate inputs x_{lm}^{ij} and input aggregator c_m^{ij}. Intermediate inputs x_{lm}^{ij} among 13 industries and 9 state groups from 1977 to 2007 were derived from the regional econometric input–output model, as discussed in Chap. 2. As given by Eq. (7.5), the input aggregator c_{mt}^{ij} can be calculated by summing up all intermediate inputs from sector i at all locations l to sector j at location m at time t,

$$c_{mt}^{ij} = \sum_l x_{lmt}^{ij}. \tag{7.13}$$

7.3.1.1 Share Parameter Calibration

Because the number of intermediate inputs from industries and locations entering (7.12) and (7.11) is large, we make the theoretical assumption, common in the New Economic Geography literature, that the value of the share parameter ε_{lm}^{ij} is the sample period average of location l's share of all intermediate input i used in the production of output j at location m arriving from all locations. This parameter's value is calculated from original intermediate input data using following formula:

$$\varepsilon_{lm}^{ij} \equiv \frac{\sum_t x_{lmt}^{ij}}{\sum_l \sum_t x_{lmt}^{ij}}. \tag{7.14}$$

7.4 Estimation of Dynamic Equations for Capital, Labor, and Input Demands

To estimate the parameters of the dynamic equations for capital, labor, and intermediate input demands, we have adopted a two-step strategy. We first estimate simultaneously the capital and labor equations for each sector and state combination, j and m, of which there are 127 in all, to obtain estimates of the returns-to-scale and elasticity parameters, κ_m^j and ρ_m^j. Then, taking these parameter estimates as given, we estimate the 13-equation blocs of demand equations for intermediate inputs in which these parameters feature. (In the second step, we also employ the calibrated values of the share parameters in the first-stage Cobb-Douglas aggregator function, discussed above, ε_{lm}^{ij}.) There are $13 \times 13 \times 9 \times 9 = 13{,}689$ such blocs in all, separated on the basis of $i - j - m$ pairs, that is, by allowing for variation in l and t. We find that this strategy reduces the overall time required for maximum-likelihood estimation when there are many cross-equation restrictions on parameters. Estimation of the stochastic differential equation systems is carried out by using Wymer's (1993) quasi-FIML nonlinear continuous-time estimator as implemented in the program ESCONA in the WYSEA package (Wymer 2006). Because the model is highly nonlinear in the parameters and maximum likelihood estimation is sensitive to misspecification, we have imposed bounds on the values that parameter estimates can assume (see Tables 7.14 and 7.15) that are in keeping with values reported in macro-econometric literatures.

7.4.1 Step 1: Capital and Labor Equations

Table 7.9 provides the details of Step 1 of the estimation process. The disequilibrium-adjustment equations for capital and labor are similar to those specified in Chap. 5 but are in terms of natural logarithms. They are:

$$\dot{lnK}_m^j = \upsilon_m^{Kj}\left(lnK_m^{j\,*} - ln\,K_m^j\right), \tag{7.15}$$

$$\dot{lnL}_m^j = \upsilon_m^{Lj}\left(lnL_m^{j\,*} - lnL_m^j\right), \tag{7.16}$$

where

$$K_m^{j^*} = \left[\frac{\alpha_m^{Kj}}{\pi_m^j}\kappa_m^j\frac{P_m^j}{ucc_m^j}\right]^{1/\left(1+\rho_m^j\right)} X_m^{j\,\left(\kappa_m^j+\rho_m^j\right)/\left(\kappa_m^j+\kappa_m^j\rho_m^j\right)}, \text{ and}$$

$$L_m^{j^*} = \left[\frac{\alpha_m^{Lj}}{\pi_m^j}\kappa_m^j e^{-\rho_m^j\lambda_m^j\cdot t}\frac{P_m^j}{W_m^j}\right]^{1/\left(1+\rho_m^j\right)} X_m^{j\,\left(\kappa_m^j+\rho_m^j\right)/\left(\kappa_m^j+\kappa_m^j\rho_m^j\right)}.$$

Table 7.9 Step 1 Estimation

Type	Variable and parameter	Description	Units/Constraints
Endogenous	K_m^j	Fixed assets in real-value terms	Millions of 2001 USD
	L_m^j	Employment	Thousand person
Exogenous	X_m^j	Aggregate output in real-value terms	Millions of 2001 USD
	c_m^{ij}	Input bundles used in production	Millions of 2001 USD
	w_m^j	Nominal wage index	Unit-free (2001=1)
	P_m^j	GDP deflator	Unit-free (2001=1)
	ucc_m^j t	User cost of capital Time trend	Unit-free
Estimated parameters	κ_m^j	Returns-to-scale parameter	[1, 2]
	ρ_m^j	Substitution parameter	[1.5, 5]
	υ_m^{Kj}	Disequilibrium adjustment parameter of K_m^j	[0.1, 2]
	υ_m^{Lj}	Disequilibrium adjustment parameter of L_m^j	[0.1, 2]
	α_m^{Kj}	Factor intensity parameter of K_m^j	
	α_m^{Lj}	Factor intensity parameter of L_m^j	
	λ_m^j	Labor augmenting rate of technical progress	
Constant	π_m^j	Mark-up[a]	

[a]We do not estimate the mark-up in the case of each regional industry. Rather we assume that it is approximately 1.2, or that there is on average a 20% mark-up over costs

7.4.2 *Step 2: Intermediate-Input Equations*

Table 7.10 provides the details of Step 2 of the estimation process. The disequilibrium-adjustment equations for intermediate inputs are modified versions of the specification in Chap. 6 (the variables are in natural logarithms).

$$\dot{lnx}_{lm}^{ij} = \upsilon_{lm}^{ij}\left(ln\, x_{lm}^{ij\ *} - lnx_{lm}^{ij}\right). \tag{7.17}$$

where

$$x_{lm}^{ij\ *} = \frac{\varepsilon_{lm}^{ij}\alpha_m^{ij}}{\pi_m^j}\kappa_m^j\frac{P_m^j}{p_{lm}^i}X_m^{j\ \left(\rho_m^j+\kappa_m^j\right)/\kappa_m^j}c_m^{ij\ -\rho_m^j}.$$

Table 7.10 Step 2 Estimation

Type	Variable and parameter	Description	Units/Constraints
Endogenous	x^{ij}_{lm}	Intermediate inputs in real-value terms	Millions of 2001 USD
Exogenous	X^{j}_{m}	Aggregate output in real-value terms	Millions of 2001 USD
	c^{ij}_{m}	Input bundles used in production	Millions of 2001 USD
	P^{j}_{m}	GDP deflator	Unit-free (2001 = 1)
	p^{i}_{lm}	Delivered prices	Unit-free (2001 = 1)
Parameter	υ^{ij}_{lm}	Disequilibrium adjustment parameter of x^{ij}_{lm}	[0.1, 2]
	α^{ij}_{m}	Factor intensity parameter of c^{ij}_{m}	
Constant	κ^{j}_{m}	Returns-to-scale parameter	
	ρ^{j}_{m}	Substitution parameter	
	ε^{ij}_{lm}	Share parameter	
	π^{j}_{m}	Mark-up	

7.4.3 Estimation Results

Because the variable observations are in natural logarithms, the standard deviations of the error terms are Root Means Square Errors (RMSE). So the quantities in line two of Table 7.11 convey the average value of the RMSEs for the first step in the estimation process. This is also the case in Table 7.12.

As can be seen from the estimation errors reported in Tables 7.11 and 7.12, both for the differential equations and their solution levels (structural errors), the fit of the model to the data is extremely good. (In these tables, “min” refers to the lowest negative number and “max” refers to the highest positive number.) Estimates of key parameters of the capital and labor equations are presented in Tables 7.13, 7.14, 7.15, 7.16, 7.17, 7.18, 7.19, and 7.20. Estimates of key parameters in the intermediate input equations are tabulated in Appendix 7.3 (Data 2 in Electronic Supplementary Material).

7.4.3.1 Discussion of the Parameter Estimates

In the estimation, we constrain the returns-to-scale parameter to range between 1.0 and 2.0. A value greater than 1 suggests increasing return to scale (IRS). 75 of 128 estimates of this parameter are discernible from zero at a conventional level of statistical significance (statistically discernible). The estimates suggest that there was increasing returns to scale in 85 of the industrial sectors studied over the sample

Table 7.11 Estimation errors for capital and labor equations

	Error in estimated differential equation		Integral of structural errors	
	K	*L*	*K*	*L*
Mean	0.000000	−0.000001	0.000000	−0.000001
Std	0.000004	0.000004	0.000005	0.000009
Min	−0.000025	−0.000032	−0.000029	−0.000086
25%	−0.000001	−0.000001	−0.000001	−0.000001
50%	0.000000	0.000000	0.000000	0.000000
75%	0.000000	0.000001	0.000001	0.000001
Max	0.000017	0.000009	0.000023	0.000016

period. While less than half of the state groupings experienced IRS in their agricultural, mining, chemical manufacturing, and machine manufacturing sectors, seven of the nine state groupings experienced IRS in construction, food production, primary metal manufacturing, fabricated metal manufacturing, and other nondurable goods manufacturing. And all nine of nine state groupings experienced IRS in computer and electronic product manufacturing, transportation equipment manufacturing, other durable goods manufacturing, and TCU, services and government enterprises. This finding would suggest the importance of allowing for the presence of IRS in modeling changing patterns of commodity flows in a spatial economy.

We constrain estimates of the substitution parameter to range between 1.5 and 5.0, which implies values of the elasticity of substitution ranging between 0.167 and 0.4. 96 of 128 estimates of the substitution parameter are statistically discernible. The estimates suggest that there was greater substitutability between productive factors in construction, chemical manufacturing, transportation equipment, and TCU, services, and government enterprises than in other sectors. (Note that because the aggregator for intermediate inputs of the same industry but from different locations is Cobb-Douglas, the implied "Armington elasticities" would be 1.0.) This finding also would suggest the need to move beyond fixed-coefficient technologies in modeling commodity flows in a spatial economy.

The disequilibrium adjustment (or elasticity of response) parameters for capital, labor, and intermediate inputs are all constrained to range between 0.1 and 2.0. The absolute value of the inverse of these parameter estimates provides an estimate of the so-called "mean time lag," or the time required to eliminate 63% of the difference between the actual level of a variable and its partial equilibrium level (Gandolfo 1981).

Only 4 of 127 adjustment parameters for capital are statistically discernible.[10] If taken at face value, and generally speaking, food production, primary metal manufacturing, and other nondurables adjust more rapidly than the other sectors, but there is quite a bit of variability between state groupings.

[10]It may be the case that a second-order differential equation characterizing adjustment in the time rate of capital formation may be a more appropriate specification to employ.

Table 7.12 Estimation errors for intermediate input equations

	x1lm	x2lm	x3lm	x4lm	x5lm	x6lm	x7lm	x8lm	x9lm	x10lm	x11lm	x12lm	x13lm
	Error in estimated differential equations												
mean	0.0000	0.0000	0.0000	0.0000	0.0000	0.0000	0.0000	0.0000	0.0000	0.0000	0.0000	0.0000	0.0000
std	0.0002	0.0003	0.0002	0.0001	0.0001	0.0002	0.0002	0.0001	0.0005	0.0000	0.0001	0.0001	0.0004
min	−0.0025	−0.0027	−0.0018	−0.0020	−0.0019	−0.0027	−0.0029	−0.0013	−0.0150	−0.0013	−0.0021	−0.0020	−0.0043
25%	0.0000	0.0000	0.0000	0.0000	0.0000	0.0000	0.0000	0.0000	0.0000	0.0000	0.0000	0.0000	0.0000
50%	0.0000	0.0000	0.0000	0.0000	0.0000	0.0000	0.0000	0.0000	0.0000	0.0000	0.0000	0.0000	0.0000
75%	0.0000	0.0000	0.0000	0.0000	0.0000	0.0000	0.0000	0.0000	0.0000	0.0000	0.0000	0.0000	0.0000
max	0.0047	0.0093	0.0049	0.0039	0.0033	0.0049	0.0067	0.0024	0.0033	0.0002	0.0038	0.0002	0.0120
	Integral of structural errors												
mean	0.0000	0.0000	0.0000	0.0000	0.0000	0.0000	0.0000	0.0000	0.0000	0.0000	0.0000	0.0000	0.0000
std	0.0002	0.0003	0.0002	0.0002	0.0001	0.0002	0.0003	0.0001	0.0006	0.0000	0.0001	0.0001	0.0004
min	−0.0028	−0.0030	−0.0020	−0.0023	−0.0021	−0.0030	−0.0033	−0.0015	−0.0195	−0.0014	−0.0023	−0.0022	−0.0048
25%	0.0000	0.0000	0.0000	0.0000	0.0000	0.0000	0.0000	0.0000	0.0000	0.0000	0.0000	0.0000	0.0000
50%	0.0000	0.0000	0.0000	0.0000	0.0000	0.0000	0.0000	0.0000	0.0000	0.0000	0.0000	0.0000	0.0000
75%	0.0000	0.0000	0.0000	0.0000	0.0000	0.0000	0.0000	0.0000	0.0000	0.0000	0.0000	0.0000	0.0000
max	0.0052	0.0102	0.0057	0.0043	0.0036	0.0054	0.0078	0.0027	0.0042	0.0003	0.0042	0.0002	0.0133

Table 7.13 Estimates of the returns-to-scale parameter κ_m^j

	RUS	ILMIWI	INOH	PA	NJ	NY	CT	MARI	MEVT	Estimates >1
ag	1.000	1.202	1.182	1.070	1.000	1.000	1.000	1.000	1.005	4
mi	1.000	1.000	1.000	1.000	1.000	2.000	2.000	1.000	1.000	2
cs	1.094	1.000	1.000	1.002	1.066	1.050	1.314	1.023	1.184	7
fd	1.041	1.008	1.058	1.000	1.000	1.043	2.000	1.159	1.111	7
ch	1.000	1.040	1.000	1.006	1.000	1.000	1.000	1.012	1.176	4
pm	1.415	1.347	1.338	1.120	1.000	1.023	1.000	1.005	1.312	7
fb	1.178	1.200	1.147	1.114	1.000	1.029	1.546	1.000	1.512	7
ma	1.000	1.000	1.000	1.000	2.000	2.000	1.669	2.000	1.000	4
cm	1.548	2.000	2.000	2.000	2.000	2.000	2.000	2.000	2.000	9
te	1.283	1.989	1.409	1.655	2.000	2.000	2.000	2.000	1.880	9
on	1.102	1.227	1.234	1.126	1.039	1.000	1.360	1.156	1.000	7
od	1.304	1.471	1.546	1.346	1.287	1.115	1.191	1.093	1.329	9
gv	1.582	1.690	1.773	1.593	1.378	1.480	1.316	1.485	1.551	9
Estimates >1	9	10	9	10	7	10	10	10	10	85

Number of estimates with *t* ratios (The "*t* ratios" referred to are the ratio of a parameter estimate to the estimate of its asymptotic standard error. These ratios converge asymptotically to the standard *t* statistic distribution (see Gandolfo 1981) statistically discernible from zero: [1.96, 2.58): 4; [2.58, 3.29): 3; [3.29, ∞): 68

Table 7.14 Substitution parameter ρ_m^j

	RUS	ILMIWI	INOH	PA	NJ	NY	CT	MARI	MEVT
ag	5.000	4.969	5.000	5.000	5.000	5.000	5.000	5.000	5.000
mi	1.504	5.000	5.000	5.000	1.500	5.000	5.000	1.500	1.500
cs	1.500	1.500	1.500	1.500	1.500	1.500	1.500	1.500	1.500
fd	5.000	5.000	5.000	5.000	3.010	5.000	1.500	5.000	5.000
ch	1.500	1.500	1.500	1.500	1.500	1.500	1.500	1.500	2.262
pm	5.000	5.000	5.000	5.000	4.999	5.000	4.979	5.000	1.500
fb	5.000	5.000	4.870	5.000	1.500	3.156	1.500	1.795	1.500
ma	5.000	5.000	5.000	5.000	4.999	5.000	4.993	5.000	5.000
cm	5.000	5.000	5.000	5.000	5.000	5.000	5.000	5.000	5.000
te	1.500	1.600	2.036	1.500	1.635	1.500	1.500	5.000	1.500
on	5.000	5.000	5.000	5.000	5.000	4.883	5.000	4.967	4.660
od	5.000	3.357	2.450	5.000	5.000	5.000	5.000	5.000	5.000
gv	1.500	1.500	1.500	1.500	1.500	1.500	5.000	1.500	1.500

Numbers of estimates with t ratios statistically discernible from zero: [1.96, 2.58): 0; [2.58, 3.29): 2; [3.29, ∞): 94

Table 7.15 Elasticity of substitution $\sigma_m^j = \frac{1}{1+\rho_m^j}$

	RUS	ILMIWI	INOH	PA	NJ	NY	CT	MARI	MEVT
ag	0.167	0.168	0.167	0.167	0.167	0.167	0.167	0.167	0.167
mi	0.399	0.167	0.167	0.167	0.400	0.167	0.167	0.400	0.400
cs	0.400	0.400	0.400	0.400	0.400	0.400	0.400	0.400	0.400
fd	0.167	0.167	0.167	0.167	0.249	0.167	0.400	0.167	0.167
ch	0.400	0.400	0.400	0.400	0.400	0.400	0.400	0.400	0.307
pm	0.167	0.167	0.167	0.167	0.167	0.167	0.167	0.167	0.400
fb	0.167	0.167	0.170	0.167	0.400	0.241	0.400	0.358	0.400
ma	0.167	0.167	0.167	0.167	0.167	0.167	0.167	0.167	0.167
cm	0.167	0.167	0.167	0.167	0.167	0.167	0.167	0.167	0.167
te	0.400	0.385	0.329	0.400	0.379	0.400	0.400	0.167	0.400
on	0.167	0.167	0.167	0.167	0.167	0.170	0.167	0.168	0.177
od	0.167	0.230	0.290	0.167	0.167	0.167	0.167	0.167	0.167
gv	0.400	0.400	0.400	0.400	0.400	0.400	0.167	0.400	0.400

Only 35 of 127 adjustment parameters for labor are statistically discernible. If taken at face value these estimates imply that labor markets are more sluggish than capital markets generally speaking. Markets for labor in the Construction, fabricated metals, transportation equipment, and other durables sectors would seem to adjust more quickly than others and the states of Maine and Vermont, New Jersey, and New York would appear to be more adaptive environments for labor.

In markets for intermediate inputs few parameter estimates are statistically discernible and are small in value for most sectors, implying that full adjustment to

Table 7.16 Disequilibrium adjustment parameter of capital υ_m^{Kj}

	RUS	ILMIWI	INOH	PA	NJ	NY	CT	MARI	MEVT
ag	0.485	0.590	0.603	0.660	0.272	0.202	0.214	0.376	0.265
mi	0.252	0.309	0.314	0.315	0.245	0.100	0.100	0.856	0.566
cs	0.298	0.145	0.162	0.131	0.131	0.170	0.100	0.134	0.100
fd	0.788	0.696	0.576	0.690	0.348	0.317	0.179	0.334	0.580
ch	0.418	0.430	0.416	0.457	0.327	0.337	0.404	0.427	0.463
pm	0.798	0.531	0.602	0.265	0.286	0.253	0.279	0.478	1.208
fb	0.437	0.412	0.381	0.367	0.209	0.304	0.195	0.282	0.275
ma	0.436	0.239	0.243	0.230	0.130	0.150	0.156	0.141	0.250
cm	0.519	0.374	0.218	0.245	0.117	0.216	0.192	0.185	0.335
te	0.421	0.183	0.307	0.271	0.209	0.189	0.203	0.148	0.231
on	0.391	0.478	0.409	0.271	0.251	0.265	0.228	0.240	0.257
od	0.489	0.358	0.425	0.386	0.209	0.338	0.282	0.355	0.329
gv	0.253	0.219	0.189	0.268	0.383	0.322	0.210	0.305	0.256

Numbers of estimates with *t* ratios statistically discernible from zero: [1.96, 2.58): 0; [2.58, 3.29): 0; [3.29, ∞): 4

Table 7.17 Disequilibrium adjustment parameter of labor υ_m^{Lj}

	RUS	ILMIWI	INOH	PA	NJ	NY	CT	MARI	MEVT
ag	0.114	0.100	0.100	0.100	0.339	0.381	0.271	0.222	0.387
mi	0.141	0.166	0.141	0.157	0.257	0.795	0.471	0.145	0.161
cs	0.115	0.130	0.140	0.312	0.415	0.438	0.677	0.473	0.696
fd	0.100	0.100	0.100	0.100	0.100	0.100	0.631	0.100	0.202
ch	0.114	0.215	0.100	0.362	0.100	0.170	0.135	0.100	0.482
pm	0.278	0.238	0.258	0.252	0.100	0.173	0.103	0.137	0.100
fb	0.828	1.161	0.888	0.583	0.483	0.517	0.327	0.428	2.000
ma	0.100	0.100	0.100	0.145	0.224	0.276	0.180	0.526	0.100
cm	0.100	0.214	0.190	0.116	0.100	0.100	0.100	0.100	0.100
te	0.392	0.999	0.829	0.576	1.049	0.557	0.100	0.857	0.100
on	0.100	0.100	0.100	0.100	0.100	0.100	0.100	0.100	0.100
od	0.614	0.714	0.775	0.475	0.778	0.410	0.278	0.466	0.551
gv	0.222	0.466	0.603	0.291	0.115	0.100	0.470	0.334	0.548

Numbers of estimates with *t* ratios statistically discernible from zero: [1.96, 2.58): 2; [2.58, 3.29): 0; [3.29, ∞): 33

partial equilibrium values may take upwards of a decade. (These parameter estimates are provided in Appendix 7.3 (Data 2 in Electronic Supplementary Material) on the Springer website associated with this publication. Note that entries of "—" for estimates of adjustment parameters or factor intensity parameters of intermediate inputs signify that the volume of interindustry sales was essentially zero or insignificant for 1977–2007.)

Estimates of factor intensity parameters (in natural logarithms) for capital, labor, and intermediate inputs range widely in value and are for the most part not statistically discernible.

Table 7.18 Factor intensity parameter of capital $ln\,\alpha_m^{Kj}$

	RUS	ILMIWI	INOH	PA	NJ	NY	CT	MARI	MEVT
ag	−0.349	6.557	5.422	1.987	0.642	1.035	0.470	0.240	1.120
mi	1.200	5.283	5.385	6.193	0.938	28.030	23.950	−0.297	0.377
cs	−3.981	−4.533	−4.707	−4.324	−3.593	−4.242	−0.986	−4.181	−2.025
fd	−5.735	−7.057	−4.725	−7.468	−5.005	−4.266	1.686	−0.731	−3.612
ch	−2.166	−1.657	−2.191	−2.216	−1.855	−1.819	−2.363	−2.044	−0.510
pm	11.841	10.331	9.927	4.235	−2.035	0.203	−1.649	−3.181	−2.604
fb	2.174	2.709	0.684	−0.559	−2.683	−3.128	1.356	−3.260	−0.534
ma	−6.656	−4.646	−4.797	−4.524	18.253	19.586	12.928	17.688	−5.081
cm	15.926	22.222	22.602	21.348	23.656	22.551	19.832	22.688	15.202
te	−0.258	5.104	1.902	1.734	2.533	3.339	3.342	17.480	1.210
on	0.006	3.488	3.729	1.142	−2.532	−4.091	6.417	2.154	−4.215
od	7.822	6.134	4.116	6.734	5.805	0.554	1.899	−0.668	4.274
gv	8.001	7.967	8.312	6.746	4.570	6.098	17.324	5.658	5.341

Numbers of estimates with t ratios statistically discernible from zero: [1.96, 2.58): 2; [2.58, 3.29): 0; [3.29, ∞): 3

Table 7.19 Estimates of the factor intensity parameter of labor $ln\alpha_m^{Lj}$

	RUS	ILMIWI	INOH	PA	NJ	NY	CT	MARI	MEVT
ag	−32.840	−23.845	−24.879	−29.191	−34.831	−34.883	−34.712	−34.548	−34.564
mi	−13.517	−33.325	−33.764	−33.936	−13.956	−13.164	−17.170	−13.112	−13.307
cs	−7.015	−8.935	−9.068	−9.389	−8.627	−8.746	−6.679	−9.192	−7.722
fd	−28.833	−32.041	−29.589	−32.615	−22.298	−31.728	−8.191	−27.083	−28.242
ch	−14.134	−13.464	−14.115	−13.915	−14.823	−14.557	−14.010	−13.962	−15.916
pm	−18.048	−22.199	−21.755	−30.562	−38.893	−35.818	−38.775	−36.016	−12.208
fb	−18.891	−18.992	−20.807	−22.953	−11.645	−18.468	−7.822	−13.070	−8.305
ma	−31.218	−33.459	−33.202	−32.829	−12.709	−9.317	−16.799	−10.428	−33.740
cm	−13.732	−9.096	−10.784	−13.047	−16.592	−12.881	−16.189	−13.123	−16.211
te	−8.791	−5.024	−9.197	−7.846	−7.804	−6.602	−7.954	−10.547	−7.302
on	−24.645	−19.375	−19.862	−26.690	−30.729	−32.426	−20.602	−26.153	−30.602
od	−12.717	−8.823	−7.084	−14.794	−17.441	−22.194	−21.072	−23.360	−17.203
gv	−1.584	−2.278	−2.122	−3.279	−4.520	−3.676	−10.470	−4.350	−4.596

Numbers of estimates with t ratios statistically discernible from zero: [1.96, 2.58): 6; [2.58, 3.29): 5; [3.29, ∞): 17

Table 7.20 Estimates of the labor augmenting technical progress λ_m^j

	RUS	ILMIWI	INOH	PA	NJ	NY	CT	MARI	MEVT
Ag	0.041	0.050	0.041	0.040	0.003	0.000	0.015	0.017	0.000
Mi	0.109	0.069	0.058	0.012	0.121	0.047	0.057	0.085	0.022
Cs	0.079	0.109	0.117	0.121	0.121	0.133	0.113	0.128	0.117
Fd	0.080	0.075	0.054	0.085	0.100	0.052	0.082	0.032	0.061
Ch	0.131	0.125	0.132	0.131	0.134	0.136	0.179	0.101	0.084
Pm	0.034	0.032	0.044	0.028	0.060	0.008	0.035	0.058	0.166
Fb	0.053	0.053	0.056	0.055	0.077	0.063	0.050	0.072	0.066
Ma	−0.023	−0.049	−0.035	−0.025	0.022	0.023	0.001	0.028	−0.033
Cm	0.020	0.001	0.016	−0.001	0.005	−0.010	0.030	0.025	0.043
Te	0.123	0.099	0.101	0.104	0.093	0.117	0.134	0.082	0.205
On	0.126	0.105	0.094	0.098	0.118	0.120	0.094	0.104	0.118
Od	0.072	0.072	0.076	0.071	0.074	0.088	0.076	0.087	0.074
Gv	0.081	0.078	0.076	0.082	0.108	0.076	0.069	0.086	0.087

Numbers of estimates with t ratios statistically discernible from zero: [1.96, 2.58): 3; [2.58, 3.29): 0; [3.29, ∞): 1

Estimates of the labor augmenting technical progress parameter are neither statistically discernible nor credible in terms of other studies of changes in labor productivity over the period of 1977–2007.

7.5 Conclusions

This chapter has presented our efforts to estimate a continuous-time structural-equation model of commodity flows between states in the Midwest and Northeastern regions of the United States and the rest of the country. Our research has demonstrated that a model whose specification is based on New Economic Geography principles can fit the data very well and capture important stylized facts of the evolution of globalization in the sample period, 1977–2007.

While estimates of many key structural parameters were statistically discernible and plausible, and thereby provide support for explanations of the evolution of commodity flows that place emphasis on increasing returns to scale in key industries, cost savings through fragmentation of production processes and increasing intra-industry trade, declining relative costs of transport leading to increased transport intensity of production and consumption, the large number of parameter estimates that are not statistically discernible suggests that the model specification we have employed can be improved upon.

We hope that our making available the data we have used to conduct this analysis will encourage others to take up the challenge to rigorously model the evolution of commodity flows as well as the other side of trade flows, economic geography.

Appendix 7.1: Experiments with Conversion Factor and Distance

This appendix introduces the experiments of scaling schemes for conversion factors and distances. Our approach is to compare the results from different scaling schemes (Table 7.21) to the observed mode shares in 1993 provided in Ham et al. (2005).

Our benchmark is the observed values of 1993 for each sector. We calculate the mean squared error (MSE) for all sectors and find that the best choice in terms of smallest MSE is the combination: Conversion Factor (/10) + Distance (/10,000).

Table 7.21 Scaling schemes

	Original unit	Scales								
Conversion factor q_x^i	million$/ kiloton	*100	*10	1	/10	/100	/1000			
Distance d_{lm}	Miles			1	/10	/100	/1000	/10,000	/1,00,000	/10,00,000

Table 7.22 Mean squared error of scaling schemes

d_{lm} q_x^i	/10,00,000	/1,00,000	/10,000	/1000	/100	/10	1
Rail							
/1000	0.0111	0.0111	0.0120	0.0988	0.5726	0.7457	0.7564
/100	0.0105	0.0105	0.0112	0.0964	0.5714	0.7456	0.7564
/10	0.0078	0.0076	0.0065	0.0769	0.5592	0.7445	0.7564
1	0.0132	0.0129	0.0100	0.0251	0.4498	0.7305	0.7564
*10	0.0301	0.0300	0.0283	0.0118	0.1958	0.6081	0.7440
*100	0.0385	0.0385	0.0385	0.0389	0.0456	0.2754	0.6170
Truck							
/1000	0.0111	0.0111	0.0120	0.0987	0.5725	0.7455	0.7562
/100	0.0105	0.0105	0.0112	0.0964	0.5713	0.7454	0.7562
/10	0.0078	0.0076	0.0065	0.0769	0.5591	0.7443	0.7562
1	0.0132	0.0129	0.0100	0.0251	0.4497	0.7303	0.7562
*10	0.0301	0.0300	0.0283	0.0117	0.1957	0.6080	0.7452
*100	0.0423	0.0423	0.0423	0.0427	0.0500	0.2752	0.6233

Appendix 7.2: User Cost of Capital Data Constructed

Table 7.23 User cost of capital (ucc) data

Year	PPI	Interest rate	Depreciation rate	Inflation rate	ucc
1977	0.461428	7.75	0.046688	0.062083	0.028657
1978	0.496035	8.49	0.046574	0.070354	0.030317
1979	0.537130	9.28	0.046404	0.082910	0.030238
1980	0.589041	11.27	0.047064	0.090595	0.040743
1981	0.654650	13.45	0.048586	0.094399	0.058059
1982	0.710887	12.76	0.050669	0.061834	0.082772
1983	0.733958	11.18	0.051026	0.039100	0.090809
1984	0.750541	12.41	0.050864	0.036101	0.104222
1985	0.766402	10.79	0.051606	0.031648	0.097991
1986	0.782985	7.78	0.051962	0.020250	0.085746
1987	0.802451	8.59	0.052197	0.024572	0.091098
1988	0.813266	8.96	0.052582	0.035308	0.086917
1989	0.842826	8.45	0.053115	0.039201	0.082946
1990	0.872386	8.61	0.053769	0.037575	0.089240
1991	0.905552	8.14	0.055115	0.033655	0.093145
1992	0.924297	7.67	0.054450	0.022820	0.100129
1993	0.940159	6.59	0.054200	0.023707	0.090624
1994	0.958183	7.37	0.053962	0.021315	0.101899
1995	0.977650	6.88	0.055066	0.020998	0.100568
1996	0.994953	6.71	0.055261	0.018285	0.103551
1997	1.000000	6.61	0.055642	0.017328	0.104415
1998	0.992790	5.58	0.055884	0.011028	0.099930
1999	0.992069	5.87	0.056370	0.014336	0.099935
2000	0.996395	5.94	0.057325	0.022568	0.093818
2001	1.007931	5.49	0.057600	0.022454	0.090760
2002	1.005047	5.43	0.056963	0.015246	0.096501
2003	1.002163	5.30	0.055871	0.018855	0.090211
2004	1.010815	5.17	0.053831	0.026778	0.079604
2005	1.036049	5.04	0.052919	0.031010	0.074915
2006	1.049027	4.91	0.053106	0.030512	0.075209
2007	1.071377	4.84	0.054239	0.026911	0.081133

References

Gandolfo G (1981) Qualitative analysis and econometric estimation of continuous time dynamic models. North Holland, Amsterdam

Ham H, Kim TJ, Boyce D (2005) Implementation and estimation of a combined model of interregional, multimodal commodity shipments and transportation network flows. Transp Res B Methodol 39(1):74–75

Phillips PCB (1991) Error correction and long-run equilibrium in continuous time. Econometrica 59:967–980

Runde J (1998) Assessing causal explanations. Oxf Econ Pap 50:151–172

U.S. Energy Information Administration (2020) Form EIA-923, Power Plant Operations Report. EIA, Washington, DC

Wymer CR (1993) Continuous-time models in macroeconomics: specification and estimation. In: Gandolfo G (ed) Continuous-time econometrics. Chapman and Hall, London, pp 35–79

Wymer CR (2006) Systems estimation and analysis programs (WYSEA manual). Version: 10 November, 2006

Chapter 8
Projections of Atmospheric Emissions and Environmental Footprints Assuming Continued Globalization

Abstract This chapter presents projections from 2008 to 2030 of point-source emissions of three of the US EPA's criteria pollutants—carbon monoxide (CO), nitrogen oxide (NO_x), and sulfur dioxide (SO_2)—and volatile organic compounds (VOC) from industrial production in the nine states or multistate groupings considered in the previous chapter and nonpoint-source emissions of the criteria pollutants and VOC plus black carbon (BC) from commodity flows along the routes connecting centroids of these states and multistate groupings. This chapter also presents calculations of environmental (emissions) footprints of both industrial production and consumption (final demand) by industry, pollutant, and location. This study is one of the first to make such calculations, taking into account interindustry sales that constitute commodity flows, and demonstrates a methodology for doing so.

8.1 Introduction

As observed in the preface to this volume, one of the animating reasons to pursue the present research project was to support plausible inferences about emissions levels and air quality *if* the globalization—and growth in the transport intensity of production and consumption, hence, growth in commodity flows and the changes in economic geography they imply—experienced in the 1970s into the 2000s were to continue apace. When designing the research presented in this volume we had aspired to conduct out-of-sample simulations with a dynamic game model of shippers and carriers, as sketched in Chap. 6, comprising all the equations whose estimation was discussed in Chap. 7, under various scenarios. As also observed in the preface, such an undertaking has proven to be prohibitive at the time of this

The lead researcher for this chapter was Ziye Zhang. Valuable assistance was provided by Arash Beheshtian.

Supplementary Information The online version of this chapter (https://doi.org/10.1007/978-3-030-78555-0_8) contains supplementary material, which is available to authorized users.

K. Donaghy et al., *The Co-evolution of Commodity Flows, Economic Geography, and Emissions*, Advances in Spatial Science,
https://doi.org/10.1007/978-3-030-78555-0_8

volume's preparation, given the challenges of the tasks involved. Nonetheless, we believe that we can convey a sense of what our funding agency, the US EPA, sought to learn about emissions and air quality, assuming continued globalization, by conducting exercises based upon extrapolation of trends in commodity flows beyond the sample period, 1977 to 2007, on which our econometric work has been based.

In this chapter, we present projections from 2008 to 2030 of point-source emissions of three of the US EPA's criteria pollutants—carbon monoxide (CO), nitrogen oxide (NO_x), and sulfur dioxide (SO_2)—and volatile organic compounds (VOC) from industrial production in the nine states or multistate groupings considered in the previous chapter and nonpoint-source emissions of the criteria pollutants and VOC plus black carbon (BC) from commodity flows along the routes connecting centroids of these states and multi-state groupings. In addition, we present calculations of the environmental (or emissions) footprints by industry, pollutant, and location for both industrial production and consumption (final demand). Unlike in Chap. 5, we do not characterize how these emissions would circulate through the atmosphere of the Midwestern and Northeastern states.

The next section of this chapter reviews previous related studies and identifies the gap this research fills. The third section sets out the research design we have followed, the fourth defines notation employed in the formal analysis, the fifth elaborates the computational formulae used to project emissions, and the sixth presents the industrial classification scheme we have adopted. The chapter's seventh section presents the details of the stylized transportation network we have employed to characterize commodity flows and the emission intensity factors for point-source and nonpoint-source emissions and their projected rates of change. In the eighth section, we present plots of industrial point-source emissions by location and pollutant in absolute levels and relative to 2008 levels and nonpoint-source emissions by transportation network link (or corridor) and pollutant. In the following section we discuss these plots. In the chapter's tenth section we present and discuss the environmental footprints of pollutant emissions from production by industry and location and the environmental footprints of pollutant emissions from consumption by industry and location in absolute levels and relative to 2008 levels. Plots of the environmental footprints are provided in two appendices, 8.1 and 8.2. In the chapter's tenth section we offer concluding observations and suggestions for further research. The Python code used to perform the computations whose results are presented in this chapter is provided on a website of the publisher.

8.2 Literature Review

Projections of point-source industrial emissions for the Midwestern US and the Chicago metropolitan region, based on out-of-sample solutions of regional econometric input–output (REIM) models, have been published in Tao et al. (2007) and Donaghy et al. (2017). And a structural decomposition analysis of declining point-source emissions in the face of expanding production has been published in Tao et al.

(2010). Estimates of nonpoint-source emissions of black carbon associated with freight movement through the Midwestern and Northeastern US, based on time series of commodity flows derived from a REIM, have been presented in Chap. 4 and published in Brown-Steiner et al. (2016). A concept central to these studies is that of an emissions intensity factor (EMI), which, for point-source emissions, indicates the amount of emissions in tons per million dollars of production by a particular industry, and, for nonpoint-source emissions, indicates the amount of emissions in grams (which can be aggregated to tons) per ton-mile of freight hauled by a particular mode of transport over a mile. If emissions intensity changes over time, say by a trend rate of *r* percent per year, the emissions intensity factor for some year *t* in the future, EMI_t, can be projected to be:

$$EMI_t = EMI_0 \times (1.0 + r/100)^t, \tag{8.1}$$

and the volume of emissions over a year can be projected as the product of the appropriate emissions intensity factor for a sector's industrial production over that interval or an amount of freight hauled over a particular distance in a given year.

Using the Chicago REIM, or CREIM, a 53-sector model, Tao et al. (2007) produced projections of emissions of six US EPA criteria pollutants—carbon monoxide (CO), ammonia (NH_3), nitrogen oxide (NO_x), two sizes of particulate matter (PM10 and PM2.5), sulfur dioxide (SO_2), and volatile organic compounds (VOC) for scenarios generated by the US EPA's economic growth analysis system (EGAS) that could be used to extend the 1999 National Emission Inventory (NEI99) of the US EPA (1999). The process involved (1) mapping NEI99 source classification code-based emissions into the sectoral or SIC- and NAICS-based representation used by the REIM, (2) developing sectoral EMIs corresponding to the model's sectoral scheme, (3) using the resulting EMIs with the REIM and projected activities giving rise to emissions to project future emissions, and (4) mapping the emissions back to the original NEI99 format. Making projections from 1999 to 2030, the authors of the study were able to project nuanced emissions profiles of the 53 industries modeled and concluded that, if technical change trends in the Chicago economy affecting emissions continued, it would likely be the case that emissions of all pollutants would decrease appreciably, even if economic growth continued at its historical pace.

Between 1970 and 2000, the economic output of the US Midwestern states doubled but emissions of criteria pollutants CO, NO_x, SO_2, and VOC all declined. The purpose of the study by Tao et al. (2010) was to examine the causes of this reduction in emissions. The authors employed a continuous-time REIM encompassing the economies of the five states of Illinois, Indiana, Michigan, Ohio, and Wisconsin and covering 13 industrial sectors and government. The authors mapped source classification codes of the NEI99 into the SIC and NAICS codes and then conducted a historical structural decomposition analysis (SDA), employing a revised Laspeyres decomposition method, to analyze the impacts of changes in technology, economic structure, and economic production on emissions of pollutants over the 30-year period. Among the structural changes experienced by the Midwestern economy were contractions in primary metals and fabricated metals manufacturing and expansions in electronic and other electric equipment and transportation equipment manufacturing. Food and kindred products production, chemical manufacturing, industrial manufacturing, and other nondurable and other durable manufacturing maintained steady production levels. Regarding changes in technology and production, the EMIs for mining evolved in an inverted U-shaped pattern, as the energy industry responded to the petroleum price shocks in the middle of the period, and the EMIs for agriculture and construction increased steadily. But the EMIs for the remaining 11 sectors steadily declined. The authors conjectured that technical change played a large role in the decline in point-source emissions of pollutants over the 30-year period; but, over the same period, globalization led to production activity in the United States shifting to "cleaner" sectors and a higher concentration of cleaner producing industries in the Midwest. The authors suggested that technical change leading to cleaner production accounted for about 75% of the reduction in EMIs and structural change accounted for the remaining 25%.

Donaghy et al. (2017) studied structural changes in the Chicago economy with a continuous-time REIM comprising 49 industrial sectors. The model supported analyses at more finely disaggregated temporal levels than those discussed above as well as examinations of economic and environmental implications of changes exogenous to the economy. The model's solution yielded estimates of point-source emission inventories that can be used to draw inferences about both structural economic changes and policy restrictions. The model was estimated with data from 1969 to 2000 and the estimated model fits the data very well. The seven pollutants represented in Tao et al. (2007) were also considered in this study, and, as in Tao et al. (2010), a structural decomposition analysis was conducted—in this case to examine differences between the effects of increased industrial production and increasing development and adoption of cleaner production technologies in explaining changes in patterns of emissions, both increases and decreases. The authors concluded that in some industries the production effect has dominated the technological effect and in other industries the reverse pattern has been obtained.

Because none of the studies reviewed above—excluding Chap. 4 and Brownsteiner et al. (2016)—considers *nonpoint-source* emissions, none explicitly examined the growing importance of supply chains, the increasing transport intensity of production and intra-industry trade, and distribution of goods and services in emissions accounting. We note here, however, that there are other trajectories of research complementary to commodity flow modeling and emissions projections at the sectoral level that do focus on supply chain management of *individual firms*. Much of this research has been conducted at MIT's Center for Transportation and Logistics and at the Smart Freight Centre in Amsterdam. There are many case studies of firms engaged in reducing emissions throughout their supply chains available on the websites of these organizations.[1] While these studies focus on individual firms, they fail to provide a larger-frame picture of the coevolution of commodity flows, economic geography, and emissions on a regional scale.

Lacking to this point have been studies that project into the future by industry and location both point-source and nonpoint-source emissions that are associated with industrial production and final demand and that explicitly take into account interindustry sales that makeup commodity flows so that plausible assessments of environmental footprints of production *and* consumption can be made. This study begins the task of making such projections and demonstrating a methodology for doing so, albeit at high levels of sectoral and geopolitical aggregation.

8.3 Research Design

In this study, we are extrapolating into the future commodity flows and EMIs to develop profiles of future point-source and nonpoint-source emissions of four of the US EPA's criteria pollutants and VOC. We proceed by assuming that inter-industry sales will continue to evolve until 2030 according to their trended growth or decline between 1997 and 2007. We make similar assumptions about future changes in EMIs, that they will reflect technical changes over the same period of time and will persist in doing so until 2030.[2]

Based on these assumptions, we proceed to project point-source emissions by location, industry, and by pollutant both in the case of continued historical technical progress leading to cleaner production and in the case of no technical progress.[3] We also present the projections relative to the 2008 levels. We project nonpoint-source emissions by transportation link corridor by pollutant with and without continued historical technical progress.

We then proceed to compute the total environmental footprint for production by sector and location and relative 2008 levels. We do the same for the total environmental footprint for consumption by sector and location and relative to 2008 levels.

8.4 Definitions of Variables and Parameters

The following notation is employed in the analysis discussed below:

[1] https://ctl.mit.edu/research/current-projects/sustainable-supply-chains; https://www.smartfreightcentre.org/en/

[2] We acknowledge that this exercise fails to take into account the global financial crisis of 2008 or the COVID-19 pandemic of 2020. But this is an exercise in extrapolation, not forecasting.

[3] Hence, we ae engaged in only a partial structural decomposition analysis because we are not considering the case where industrial production is held constant at 2007 levels.

X^j_{mt} is aggregate output in value terms by sector j at location m at time t.
x^{ij}_{lmt} is the shipment in value terms of commodities from industry i at location l to industry j at location m at time t.
FD^i_{lt} is the *total* final demand in value terms for the output of industry i at location l at time t from all locations m.
FD^i_{lmt} is the final demand in value terms for the output of industry i at location l at time t from just location m.
$D^v_{lmr} \equiv \sum_a d^v_a \delta^{av}_{lmr}$, is the sum of link distances for mode v, d^v_a, along route r between location l and location m, and δ^{av}_{lmr} is 1.0 if link a is part of the route and zero otherwise.
h^{iv}_{lmrt} is the physical flow (in tons) of commodities of sector i from location l to location m by mode v along at time t. (Although we index routes here, we assume that transport occurs along the shortest-distance route.)
f^v_{at} is the total physical flow of commodities of all sectors along link a by mode v at time t.
E^k_{mt} is the aggregate amount of point-source emissions of criteria pollutant k by all industries in location m at time t.
E^k_{at} is the aggregate amount of nonpoint-source emissions of criteria pollutant k from transport of commodities along link a at time t.
q^i_x is a factor for converting quantities in value terms to tonnage in industry i.
b^x_v, b^d_v, and b^i are parameters in the logit weighting function determining shares of modal usage in shipping.
φ^{jk}_t is a coefficient (an emissions intensity factor or EMI) giving the volume in metric tons of point-source emissions of criterion pollutant k in the production of \$1 million of commodity j at time t. (We are assuming in this exercise that these coefficients apply to common industries across all locations.)
φ^{vk}_{at} is a coefficient giving the volume in grams of nonpoint-source emissions of criterion pollutant k per ton-mile of transport along link a by mode v at time t.
TEFPP^{jk}_{mt} is the sum of all emissions of type k associated with the production of output j at location m at time t (the total environmental footprint of production).
TEFPC^{jk}_{mnt} is the sum of all emissions of type k associated with consumption of output j from location m at location n at time t.
TEFPC^{jk}_{nt} is the sum of all emissions of type k associated with consumption of output j from *all* locations m at location n at time t (the total environmental footprint of consumption).

8.5 Computational Formulae for Projections

8.5.1 Commodity and Freight Flow Equations

Material Balance Equation (Imputation of Total Final Demand for Output of Industry i at Location l)

$$\mathrm{FD}^{i}_{lt} = X^{i}_{lt} - \sum_{m}\sum_{j} x^{ij}_{lmt}, \forall i, \forall l. \tag{8.2}$$

Final Demand Allocation to locations m

$$\mathrm{FD}^{i}_{lmt} = \mathrm{FD}^{i}_{lt} \cdot \sum_{j} X^{j}_{mt} \left(\sum_{v} \left(D^{v}_{lmr*} \omega^{iv}_{lm} \right) \right)^{-1} / \sum_{m}\sum_{j} X^{j}_{mt} \left(\sum_{v} \left(D^{v}_{lmr*} \omega^{iv}_{lm} \right) \right)^{-1}, \forall i, \forall l, \forall m. \tag{8.3}$$

where $\omega^{iv}_{lm} = \exp\left(b^{x}_{v} q^{i}_{x} + b^{d}_{v} D^{v}_{lmr} + b^{i} \delta^{v}\right) / \sum_{v} \exp\left(b^{x}_{v} q^{i}_{x} + b^{d}_{v} D^{v}_{lmr} + b^{i} \delta^{v}\right).$

Conservation of Flow Equation

$$\sum_{r} h^{iv}_{lmrt} = \omega^{iv}_{lm} \sum_{j} x^{ij}_{lmt} / q^{i}_{x} + \omega^{iv}_{lm}\, FD^{i}_{lmt} / q^{i}_{x}, \forall i, \forall l, \forall m, \forall v. \tag{8.4}$$

In which $\delta^{v} = 1$ if the mode of shipment is by truck and 0 otherwise, and b^{i} is the industry-specific coefficient for the indicator variable.[4]

Link Physical Flow Equation[5]

$$f^{v}_{at} = \sum_{i}\sum_{lmr} h^{iv}_{lmrt} \delta^{av}_{lmr}, \forall a, \forall v, \tag{8.5}$$

where δ^{av}_{lmr} is 1.0 if link a is part of the route and zero otherwise.

[4]Note that the logit function determining modal shares, ω^{iv}_{lm}, is *static*. We are assuming that shifts in modal shares over the last few decades between 2008 and 2030 will not be dramatic and so invalidate our projections.

[5]Note that this equation does *not* account for the *haulage of through traffic*. We are focusing just on the flow of freight deriving from the commodity flows we project. We are also implicitly assuming that the link capacities for physical flows by mode v along link a are not exceeded.

8.5.2 Point-Source and Nonpoint-Source Emissions

Point-Source Emissions of Criterion Pollutant k at Location m

$$E_{mt}^{k} = \sum_{j} \varphi_{t}^{jk} X_{mt}^{j}, \forall k, \forall m. \tag{8.6}$$

NonPoint-Source Emissions of Criterion Pollutant k along Link a by All Modes of Transport

$$E_{at}^{k} = \sum\nolimits_{v} \varphi_{at}^{vk} f_{at}^{v} d_{a}^{v}, \forall k, \forall a. \tag{8.7}$$

8.5.3 Environmental Footprint of Production

All point-source emissions of type k at all locations l associated with the production of the intermediate inputs i used in the production of output j at location m at time t are:

$$\sum_{l} \sum_{i} \varphi_{t}^{jk} x_{lmt}^{ij}, \forall j, \forall m, \forall k. \tag{8.8}$$

All point-source emissions of type k associated with production of output j at location m at time t are:

$$\varphi_{t}^{jk} X_{mt}^{j}, \quad \forall j, \forall m, \forall k. \tag{8.9}$$

All nonpoint-source emissions of type k associated with the supply of intermediate inputs i from all locations l for production of output j at location m at time t are:

$$\sum_{l} \sum_{v} \sum_{a} \varphi_{t}^{vk} d_{a}^{v} \delta_{lm}^{av} \cdot \sum_{i} \frac{\omega_{lm}^{iv} x_{lmt}^{ij}}{q_{x}^{i}}, \forall j, \forall m, \forall k. \tag{8.10}$$

where, ω_{lm}^{iv} is as defined above.

The sum of (8.8)–(8.10) gives the sum of all emissions of type k associated with the production of output j at location m at time t, TEFPP_{mt}^{jk}:

$$\text{TEFPP}_{mt}^{jk} = \sum_{l}\sum_{i}\varphi_t^{jk}x_{lmt}^{ij} + \varphi_t^{jk}X_{mt}^{j} + \sum_{l}\sum_{v}\sum_{a}\varphi_t^{vk}d_a^{v}\delta_{lm}^{av}\cdot\sum_{i}\frac{\omega_{lm}^{iv}x_{lmt}^{ij}}{q_x^{i}}, \forall j, \forall m, \forall k. \tag{8.11}$$

8.5.4 Environmental Footprint of Consumption

The share of emissions of type k associated with the production of output j at location m that will be sold to *all* sources of final demand in all locations, FD_{mt}^{j}, is:

$$\left(\text{FD}_{mt}^{j}/X_{mt}^{j}\right)\cdot\text{TEFPP}_{mt}^{jk}, \forall j, \forall m, \forall k. \tag{8.12}$$

This expression gives the sum of point-source and nonpoint-source emissions associated with the final demand for output j produced at location m at time t, FD_{mt}^{j}. Of course, this total amount of final demand will not be consumed entirely at location m. To allocate the emissions associated with this final demand level to locations where it will be consumed, we need to account for both (1) the share of final demand to be consumed at other locations and (2) the additional nonpoint-source emissions that will be incurred in transporting shares of FD_{mt}^{j} to these locations. We can account for the first part of this allocation—the share to be consumed at other locations—by employing a gravity model construct similar to the one used earlier to allocate final demand.

Let us denote by the coefficient, g_{mnt}, the fraction of final demand for output j produced at location m at time t that will be allocated to location n. (Note that m now denotes the origin location and n denotes the destination location, but all variables and parameters are interpreted as before):

$$g_{mnt} = \sum_{j}X_{nt}^{j}\left(\sum_{v}\left(D_{mnr*}^{v}\omega_{mn}^{jv}\right)\right)^{-1} \Big/ \sum_{n}\sum_{j}X_{nt}^{j}\left(\sum_{v}\left(D_{mnr*}^{v}\omega_{mn}^{jv}\right)\right)^{-1}. \tag{8.13}$$

Multiplying this coefficient by the share of emissions of type k associated with the production of output j at location m that will be sold to all sources of final demand in all locations, FD_{mt}^{j}—that is (8.12)—we obtain an estimate of the emissions associated with location n's share of final demand for the output of industry j at location m at time t *before* it is shipped to location n:

$$g_{mnt}\cdot\left(\text{FD}_{mt}^{j}/X_{mt}^{j}\right)\cdot\text{TEFPP}_{mt}^{jk}, \forall j, \forall m, \forall k. \tag{8.14}$$

To account for the additional nonpoint-source emissions associated with shipping this amount to location n we need to compute and add the following quantity to (8.14):

$$\sum_{v}\omega_{mn}^{jv}\cdot g_{mnt}\cdot\left(\mathrm{FD}_{mt}^{j}/q_{x}^{j}\right)\sum_{a}\varphi_{t}^{vk}d_{a}^{v}\delta_{mn}^{av},\forall j,\forall m,\forall k. \tag{8.15}$$

The total emissions of type k associated with consumption of output j from location m at location n at time t, $\mathrm{TEFPC}_{mnt}^{jk}$, will be the sum of (8.14) and (8.15):

$$\begin{aligned}\mathrm{TEFPC}_{mnt}^{jk} &= g_{mnt}\cdot\left(\mathrm{FD}_{mt}^{j}/X_{mt}^{j}\right)\cdot\mathrm{TEFPP}_{mt}^{jk}\\ &+\sum_{v}\omega_{mn}^{jv}\cdot g_{mnt}\cdot\left(\mathrm{FD}_{mt}^{j}/q_{x}^{j}\right)\sum_{a}\varphi_{t}^{vk}d_{a}^{v}\delta_{mn}^{av},\quad\forall j,\forall m,\forall n,\forall k.\end{aligned} \tag{8.16}$$

The total emissions of type k associated with consumption of output j from *all* locations m at location n at time t, TEFPC_{nt}^{jk}, will be:

$$\mathrm{TEFPC}_{nt}^{jk}=\sum_{m}\mathrm{TEFPC}_{mnt}^{jk},\forall j,\forall n,\forall k. \tag{8.17}$$

In calculating the estimates of pollutant emissions, we assume that industry production levels and interindustry sales, hence, commodity flows continue to evolve according to the trends observed between 1997 and 2007. Hence, we generate the aggregate output X_{lt}^{i} (or X_{mt}^{j}) and the interindustry sales x_{lmt}^{ij} in (8.2) according to the following procedure.

1. We compute trend extrapolations of intermediate input levels for 18 years forward using percentage growth rates from 1997 to 2007 to derive the extrapolations.

$$x_{lmt}^{ij}=x_{lm,t=2007}^{ij}e^{\rho_{lm}^{ij}\cdot(t-2007)},\forall t\geq 2008$$

   where $\rho_{lm}^{ij}=\left(\ln x_{lm,t=2007}^{ij}-\ln x_{lm,t=1997}^{ij}\right)/10.$

2. We use the ratios of gross output to intermediate inputs, GO/II, for 2008 and 2017 obtained from Randall Jackson's program, I–O Snap (Jackson and Court 2020) to derive proportionate rates of change in these ratios, r_{lm}^{ij}, by computing [ln (GO/II17)-ln(GO/II08)]/9:

$$GOII_{lmt}^{ij}=GOII_{lm,t=2008}^{ij}e^{r_{lm}^{ij}\cdot(t-2008)},\forall t>2008$$

   where $r_{lm}^{ij}=\left(ln\,GOII_{lm,t=2017}^{ij}-ln\,GOII_{lm,t=2008}^{ij}\right)/9.$

Table 8.1 Industrial classification scheme

Sector	Description
1	Agriculture, forestry, fishing, and hunting
2	Mining
3	Construction
4	Food production manufacturing
5	Chemical manufacturing
6	Primary metal manufacturing
7	Fabricated metal product manufacturing
8	Machinery manufacturing
9	Computer and electronic product manufacturing
10	Transportation equipment
11	Other nondurable manufacturing
12	Other durable manufacturing
13	TCU, services, and government enterprises

3. We then sum up the intermediate input sales to other sectors by each industry j in location m and multiply each sum by the GO/II ratio appropriate for the industry, location, and year to generate the relevant gross output level:

$$X^i_{lt} = GOII^{ij}_{lmt} \cdot \sum_m \sum_j x^{ij}_{lmt}, \forall t \geq 2008$$

Taking this approach to generating values of X^i_{lt} embeds sectoral technical progress endemic to the 1997–2007 time series.

8.6 Classification of Industrial Sectors

We employ the same sectoral scheme as in the previous chapter, set out in Table 8.1.

8.7 Details of Transportation Network and Emission Intensity Factors

In Table 8.2, we provide in kilometers shortest-route distances between centroids of the nine geopolitical groupings of this study and estimates of the average lengths of intrazonal trips.

In Fig. 8.3, we provide plots of nonpoint-source emissions for stylized links connecting centroids of the nine geopolitical groupings in this study. Table 8.3

Table 8.2 Distances between Centroids and Intrazonal Distances (in kilometers)

	RUS	ILMIWI	INOH	PA	NJ	NY	CT	MARI	MEVT
RUS	2200.0	2818.9	3432.5	3899.7	4126.5	4148.3	4239.9	4492.2	4717.8
ILMIWI	2818.9	300.0	519.4	1236.5	1228.3	1250.1	1341.7	1594.0	1819.6
INOH	3432.5	519.4	200.0	717.1	708.9	730.7	822.3	1074.6	1300.2
PA	3899.7	1236.5	717.1	300.0	226.8	248.6	340.2	592.5	818.1
NJ	4126.5	1228.3	708.9	226.8	100.0	21.8	113.4	365.7	591.3
NY	4148.3	1250.1	730.7	248.6	21.8	150.0	91.6	343.9	569.5
CT	4239.9	1341.7	822.3	340.2	113.4	91.6	50.0	252.3	477.9
MARI	4492.2	1594.0	1074.6	592.5	365.7	343.9	252.3	75.0	236.2
MEVT	4717.8	1819.6	1300.2	818.1	591.3	569.5	477.9	236.2	200.0

RUS rest of the United States, *ILMIWI* Illinois, Michigan, and Wisconsin, *INOH* Indiana and Ohio; *PA*, Pennsylvania; *NJ*, New Jersey; *NY*, New York; *CT*, Connecticut; *MARI*, Massachusetts and Rhode Island; *MEVT*, Maine and Vermont

Table 8.3 Stylized links connecting centroids

Link	Centroids linked
GCD_N	Denver, CO and Great Circle Distance Northern Connection Node
GCD_S	Denver, CO and Great Circle Distance Southern Connection Node
AAA	Portland, ME and Worcester, MA
BBB	Portland, ME and Hartford, CT
CCC	Worcester, MA and Hartford, CY
DDD	Hartford, CT and Albany, NY
EEE	Albany, NY and New Brunswick, NJ
FFF	New Brunswick, NJ and Harrisburg, PA
GGG	Harrisburg, PA and Columbus, OH
HHH	New Brunswick, PA and Columbus, OH
III	Columbus, OH and Chicago, IL
JJJ	Columbus, OH and Great Circle Distance Northern Connection Node
KKK	Harrisburg, PA and Great Circle Distance Southern Connection Node
LLL	Chicago, IL and Great Circle Distance Northern Connection Node

provides a key for identifying the links. We assume that both truck and rail traffic proceed along these links (or corridors).[6]

8.7.1 Emission Intensity Factors

For our projections of emissions of carbon monoxide (CO), nitrous oxide (NOx), sulfur dioxide (SO_2), volatile organic compounds (VOC), and black carbon (BC) we adopt the following EMIs, taken from multiple sources, for the base year of 2008, and assumptions about technical progress that will lead to reductions in their values. The EMIs for the three US EPA criteria pollutants and volatile organic compounds and projected rates of change (in Tables 8.4 and 8.5), used in estimating emissions from industrial production, were based on previous analysis of emissions for a sectoral scheme by Tao et al. (2010). They were updated to 2008 using rates of change assumed in that study and put in terms of constant 2001 dollars. We acknowledge that this approach may misrepresent emission profiles in non-Midwestern states, but plots providing comparisons of emissions to 2008 levels should be free of such misrepresentation. Moreover, the US EPA's National Emissions Inventory for 2008 reports similar densities of emissions throughout the 13 states involved in this study. Source data on EMIs and rates of change for nonpoint-source emissions (in Tables 8.6 and 8.7) were taken from the 2020 US DOT FHA publication, National Freight Transportation Trends and Emissions, Chap. 2, "Freight Movement and Air Quality."

[6]Ham et al. (2005) employ virtual 'spider' networks in their commodity flow model.

Table 8.4 Emission intensity factors (EMIs) in terms of tons of emissions per $Million of sectoral output in 2001 dollars in 2007

Sector	Pollutant			
	CO	NOx	SO_2	VOC
1	43.83615	9.46210	5.05447	24.39448
2	6.28343	1.35833	0.72074	3.49284
3	1.36757	0.29570	0.15709	0.70630
4	0.76821	0.19133	0.08528	0.44796
5	1.64898	0.41593	0.19189	0.95565
6	5.09751	1.28106	0.58280	2.95655
7	0.68828	0.17470	0.07818	0.39570
8	0.40866	0.09982	0.04975	0.25385
9	0.50186	0.12478	0.05686	0.29117
10	0.43017	0.10814	0.04975	0.25385
11	1.77803	0.44088	0.20611	1.03031
12	1.71350	0.43256	0.19900	0.99299
13	3.89937	0.37184	0.34402	1.89300

Table 8.5 Average annual % change in EMIs, 1970–2000

Sector	Pollutant			
	CO	NOx	SO_2	VOC
1	0	0	0	0
2	0	0	0	0
3	0	0	0	−1.17
4	−3.56	−1.49	−3.68	−3
5	−3.56	−1.49	−3.68	−3
6	−3.56	−1.49	−3.68	−3
7	−3.56	−1.49	−3.68	−3
8	−3.56	−1.49	−3.68	−3
9	−3.56	−1.49	−3.68	−3
10	−3.56	−1.49	−3.68	−3
11	−3.56	−1.49	−3.68	−3
12	−3.56	−1.49	−3.68	−3
13	−1.07	−11.94	−4.92	−2.97

Table 8.6 Coefficients of nonpoint-source emissions in grams per ton-mile in 2007

Pollutant	Rail	Truck
VOC	0.03564	0.07001
CO	0.09553	0.32957
NOx	0.65594	1.29438
SO2	0.00654	0.03921
BC	0.01421	0.01700

Table 8.7 Average annual % change in nonpoint-source emission coefficients

Pollutant:	Rail	Truck
VOC	−6.22	−6.22
CO	−5.20	−5.20
NOx	−3.94	−3.94
SO2	−2.16	−2.16
BC	−1.50	−5.90

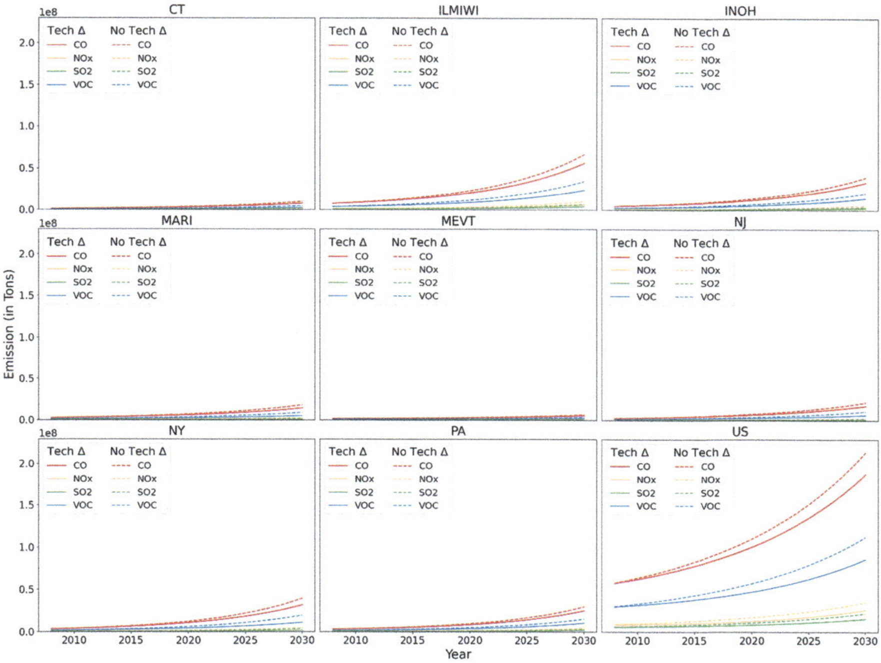

Fig. 8.1 Total industrial point-source emissions by location and pollutant

8.8 Plots of Emissions

In Fig. 8.1 are plots of point-source emissions (in tons) of five industrial pollutants projected for 2008 to 2030, given by location and pollutant under two different assumptions: (1) that technical progress leading to cleaner production continues over the projection period, and (2) that no further such technical progress is made over the projection period. Fig. 8.2 presents the same data plotted in Fig. 8.1 relative to their 2008 values. In Fig. 8.3 are plots of non-point-source emissions (in tons) projected for 2008 to 2030, given by stylized transportation network link and pollutant under the two assumptions: (1) that technical progress leading to cleaner transport continues over the projection period, and (2) that no further such technical progress is made over the projection period. Fig. 8.4 presents the same data plotted in Fig. 8.3 relative to their 2008 values.

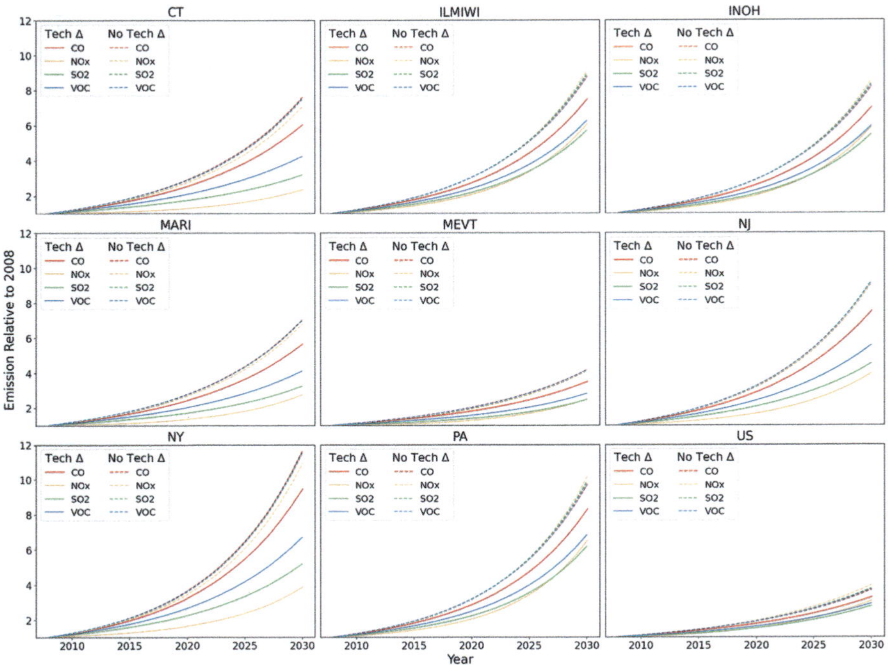

Fig. 8.2 Total industrial point-source emissions by location and pollutant relative to 2008

8.9 Discussion of Emissions Plots

The plots of point-source emissions by location and pollutant in Fig. 8.1 do not portend substantial changes in emission patterns whether or not technical change is held out. Among the Midwestern and Northeastern states or state groupings, ILLMIWI, INOH, and NY and PA are the leading polluters. RUS totals, corresponding to a larger aggregation of state economies, are higher, as one would expect. Examining projected point-source emissions relative to 2008 levels in Fig. 8.2, one can see a precipitous climb the further out in time one goes (partly a function of the scale of the plots). The plots in Fig. 8.2 suggest that (a) physical effects (the growth in output) will dominate technical effects (adoption of cleaner production technologies) in raising emissions; (b) technical changes makes a big difference (in relative terms); (c) even with growth rates in ratios, the changes are not large magnitudes; and (d) CO is the pollutant whose emissions are increasing the fastest while VOC emissions are increasing the second fastest.

The plots of nonpoint-source emissions by link and pollutant in Fig. 8.3 portend that, with the exception of KKK, emission levels will remain relatively flat, suggesting that technical changes in transport that reduce emissions are keeping pace with increasing volumes of traffic and that the absence of continued improvements in cleaner transport would make a difference. SO_2 is the nonpoint-source

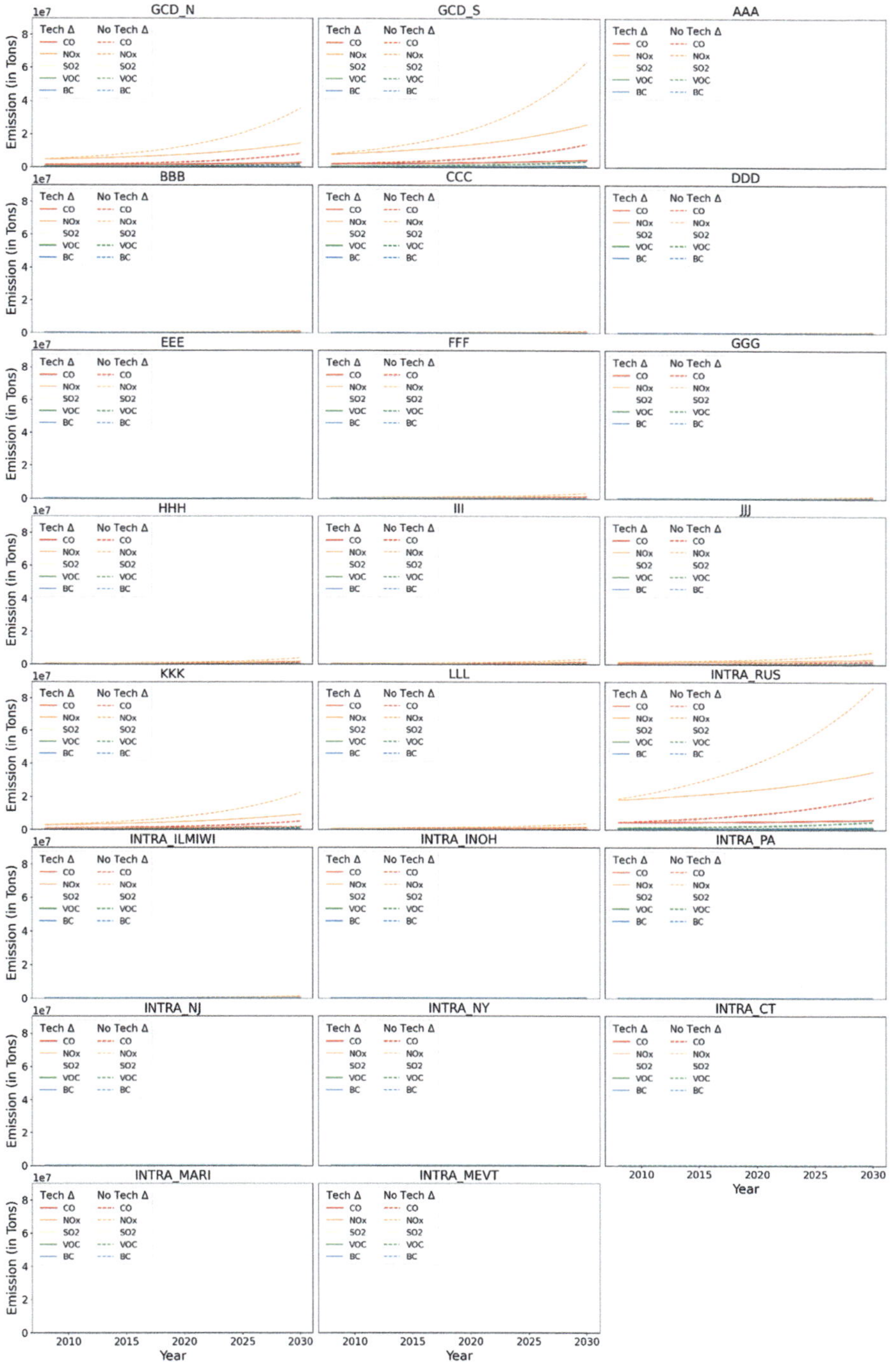

Fig. 8.3 Total nonpoint-source emissions by link and pollutant

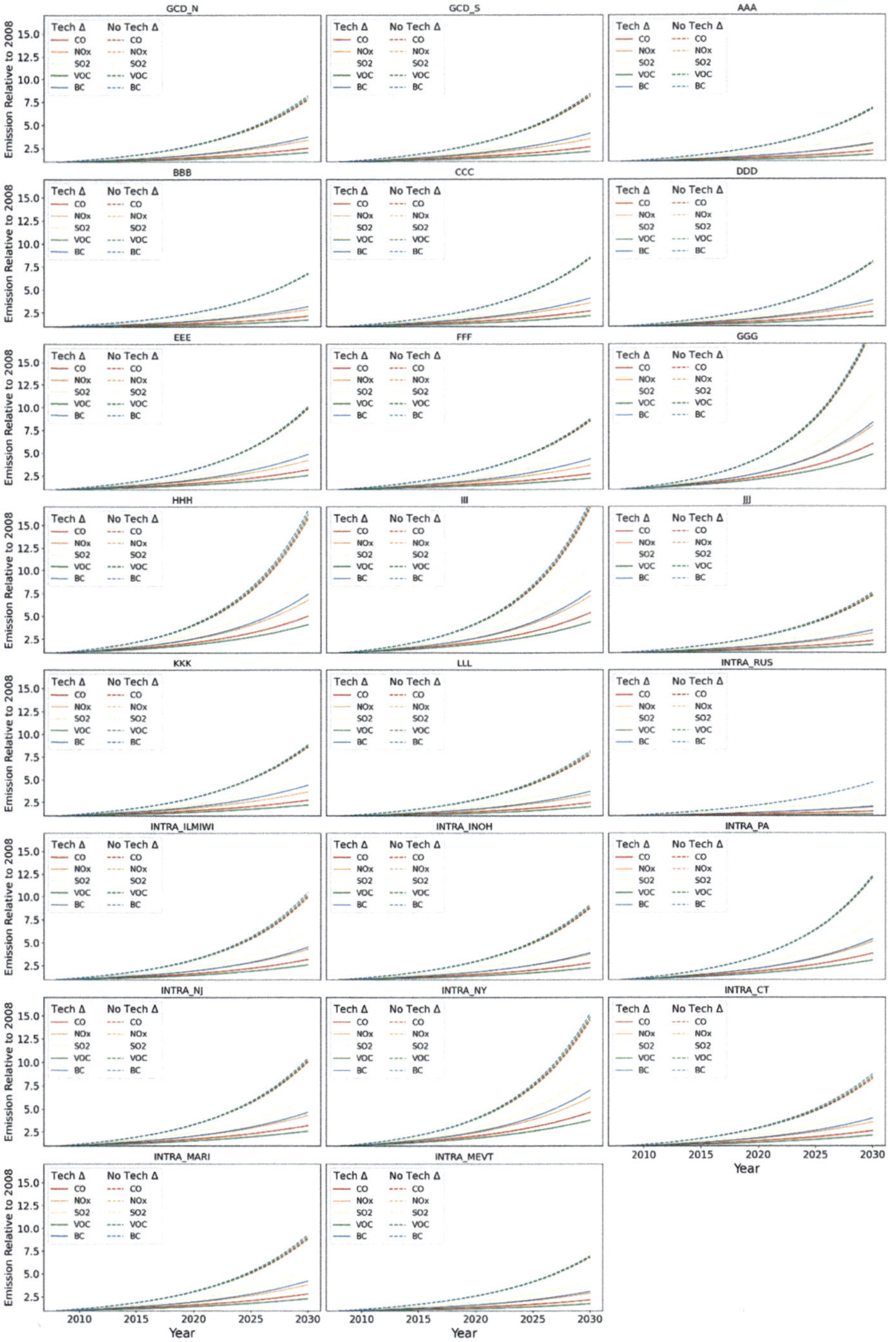

Fig. 8.4 Total nonpoint-source emissions by link and pollutant relative to 2008

pollutant whose volume is projected to increase the most. The highest volumes of projected emissions with and without technical change are associated with links GCD-N and GCD-S and intra-RUS, corresponding to a higher volume of traffic over a larger area covered. Intrazonal emissions for the eight other states or state groupings are negligible. Considered relative to 2008 levels, however, there are discernible changes going forward. The rankings of the emissions by volume are SO_2, BC, NO_x, CO, and VOC. Technical change can be seen to make a big difference resulting in a more-than-halving of relative emissions without such change. In relative terms, changes in GCD-N, GCD-S, and intra-RUS are not noticeably different from changes in other links or intrazonal routing.

8.10 Plots and Discussion of Environmental Footprints

We turn now to plots of environmental footprints of production (in Appendix 8.1). With the exception of Fig. 8.5, these footprints aggregate *both* point-source and nonpoint-source emissions attributable to locations of production and the supply chains enabling production in those locations, thereby providing a more complete composite picture of an industry's environmental effects. In Fig. 8.5, which presents plots of total BC emissions associated with production by industry and location, all emissions plotted are nonpoint-source. Exponential increases are projected for mining and construction in RUS with little change in the Midwestern and Northeastern states. There is no discernable change in production emissions attributable to agriculture, forestry, and fisheries (agriculture), although the EMIs for this sector are the largest (see Table 8.3). BC emissions related to production in most sectors of the Midwestern and Northeastern states are stable. Where changes do occur, they are increasing most noticeably for the RUS in mining and construction, less so for other nondurables, transportation equipment, TCU, services, and government enterprises (TCU), and food production, and declining in fabricated metals. Relative to 2008 levels, BC emissions attributable to production in Fig. 8.6 are sharply increasing in mining and construction and increasing moderately in TCU in all states and state groupings. They are fairly stable for agriculture, chemical products, prime metals, other nondurables, and other durables (with the exception of RUS). And production would appear to be getting cleaner in food production, fabricated metals, computers and electronics, transportation equipment, and machinery manufacturing (with the exception of RUS in the latter two).

From Fig. 8.7 one can observe that the environmental footprint for point-source and nonpoint-source CO emissions by most industries in most localities in absolute levels changes little with the exception of noticeable upturns by the sectors of agriculture, mining, construction, and TCU for RUS. There are less noticeable upturns in mining and TCU for all others. Relative to 2008 levels (Fig. 8.8), CO emissions for all localities increase noticeably in mining, construction, and TCU, and lest noticeably in agriculture and computer and electronics.

Patterns of change in NO_x emissions from production by most industries, in absolute levels and or relative to 2008 (portrayed in Figs. 8.9 and 8.10), are in most cases similar to those of CO, with the exception that there are virtually no changes in emissions in TCU.

Future SO_2 emission patterns (portrayed in Figs. 8.11 and 8.12) are also similar to those of CO with minor variations. In the case of VOC emissions (Figs. 8.13 and 8.14) there are noticeable increases in emissions of mining, construction, and less noticeable increases in agriculture, chemical products, and TCU.

Generally speaking, it would appear, from the examination of projected environmental footprints of industrial production that a continuation of trends experienced under globalization in the period of 1997 to 2007 would not lead to an increase in emissions in most industries in most locations, apart from mining and construction. Or, in other words, the production effect and the technical effect will be in balance.

We now address environmental footprints of consumption (in Appendix 8.2). Considering just the absolute levels in the case of BC emissions (Fig. 8.15), we can see that there are increases in emissions attributed to the consumption of products from mining (or mineral extraction and energy), construction and, to a lesser extent, TCU in all localities and a decrease in BC emissions in the consumption of fabricated metals products attributable to consumption in RUS. Considering projected BC emissions relative to 2008 levels (Fig. 8.16), it is clear that there would be increases in emissions due to consumption in all localities. And, interestingly, the EFP for BC of RUS seems to be rising less than the EFPs of Midwestern and Northeastern states and state groupings. The state projected to account for the largest relative increase in EFP is CT.

Examining the EFPs for CO in absolute terms (Fig. 8.17), the biggest changes are projected to result from consumption of products of agriculture, mining, construction, and TCU by households in RUS. Relative to 2008 (Fig. 8.18), there is steady growth in EFPs for agriculture, mining, construction, chemical production, transport equipment, other nondurables, other durables, and TCU with CT again increasing the most and RUS the least.

As regards NO_x, the EFPs of consumption of products from mining and construction in RUS are projected to be the largest (Fig. 8.19), whereas there is a steady increase in EFPs relative to 2008 levels for all sectors goods in all localities, with the EFPs of CT increasing the most and those of RUS the least (Fig. 8.20).

Projections of EFPs of SO_2 emissions from consumption of industrial products in absolute levels and relative to 2008 (Figs. 8.21 and 8.22) again single out mining and construction as the most important contributors. Projections of absolute levels are mostly flat for other sectors, although a faint growth effect is observed. Again the increase in emissions attributable to consumption is greatest for CT and least for RUS.

For VOC, changes in EFPs in absolute terms (Fig. 8.23) are projected to be small for most sectors except mining and TCU. Relative to 2008 (Fig. 8.24), steady but slow growth in emissions is projected and is attributable to consumption of goods produced by all sectors by households in all locations, with the exception of the consumption of products of the mining industry. Again, CT has the largest EFPs and RUS the least.

From these plots one can infer that if the developments in the spatial economies of the Midwest, Northeast and rest of the United States that occurred from 1997 to 2007 continued, then consumption patterns abetted by increasing transport intensity would steadily increase the EFPs of households.

In fact, the future of EFPs of production by the mining, gas, and energy industry and consumption of its products foretold by this exercise is not likely to occur, given recent actions taken within the industry in response to climate change and the emergence of commercially viable renewable energy sources. And we should note that trends from 1997 to 2007 predated the rise of Amazon and other major e-trading firms, which would likely increase the EFPs of consumption, so further changes in the evolution of retail logistics are not accounted for.

8.11 Concluding Observations and Future Directions

The research this chapter has presented is novel in that it bases projections of point-source and nonpoint-source emissions of various pollutants on a characterization of a spatial economy in which the flows of intermediate inputs at every stage of industrial production and final demand are taken into account.

The picture that emerges is that from the perspective of production, environmental effects (in the form of atmospheric emissions) of increases in production are being offset by improvements in technology, cleaner production—even with increasing fragmentation and intra-industry trade in intermediate inputs. From the perspective of consumption, however, the production effect would seem to dominate, if not overwhelm, the technological effect. (The role of a "structural effect," due to changes in sectoral activity, commented on in Tao et al. (2010) and in Chap. 4, is less noticeable, having occurred mostly before 1997.) The increasing demand for energy products, construction, and services (including transportation services), are noticeable from both perspectives of production and consumption.

The extrapolative and conjectural research reported in this chapter can be extended in multiple ways.

- The geographical scope of the analysis can be widened to encompass the entire US spatial economy.
- The time series on commodity flows can be extended beyond 2007 by employing the methods set out in Chap. 2.
- Dynamic programming models along the lines introduced in Chap. 5 can be operationalized to forecast the evolution of freight movement on the basis of factors not considered here and associated environmental foot prints that are affected explicitly by factors not considered here.
- Dynamic game analyses considering the interaction of shippers and carriers can be conducted to assess the impacts of potential regulation of logistics operations.
- Analyses at the level of sectors and states can be merged with those of large individual firms with extensive production networks to explore how environmental footprints can be better managed.

We must leave these suggestions for other scholars to take up at other times.

Appendix 8.1: Total Environmental Footprints of Production

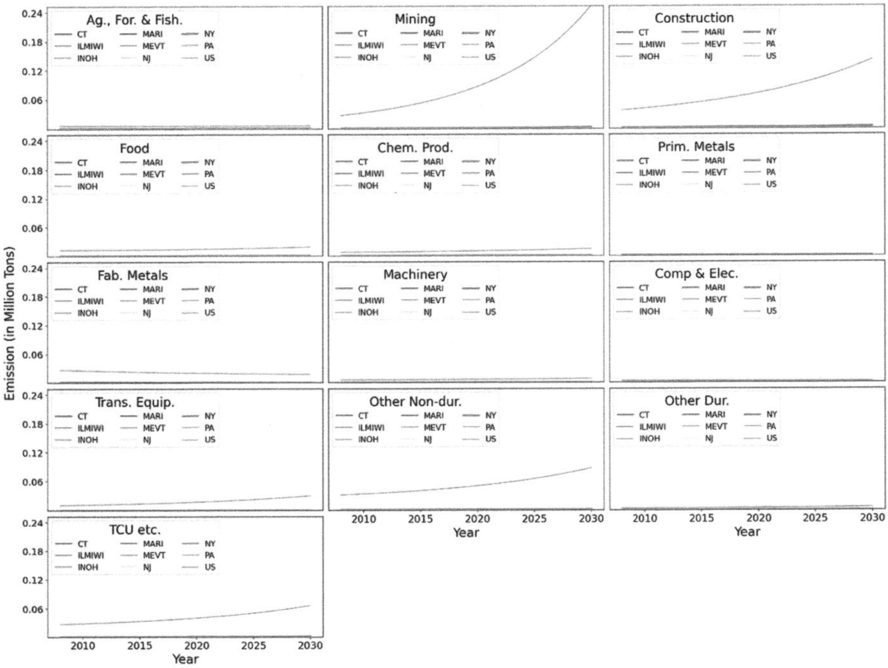

Fig. 8.5 Total environmental footprint of black carbon emissions from production by industry and location

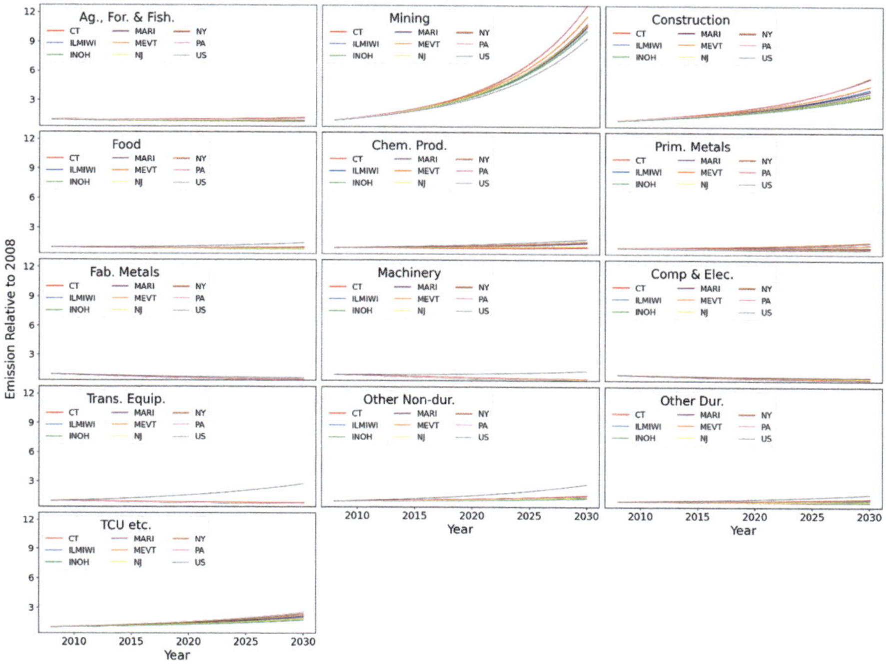

Fig. 8.6 Total environmental footprint of black carbon emissions from production by industry and location relative to 2008

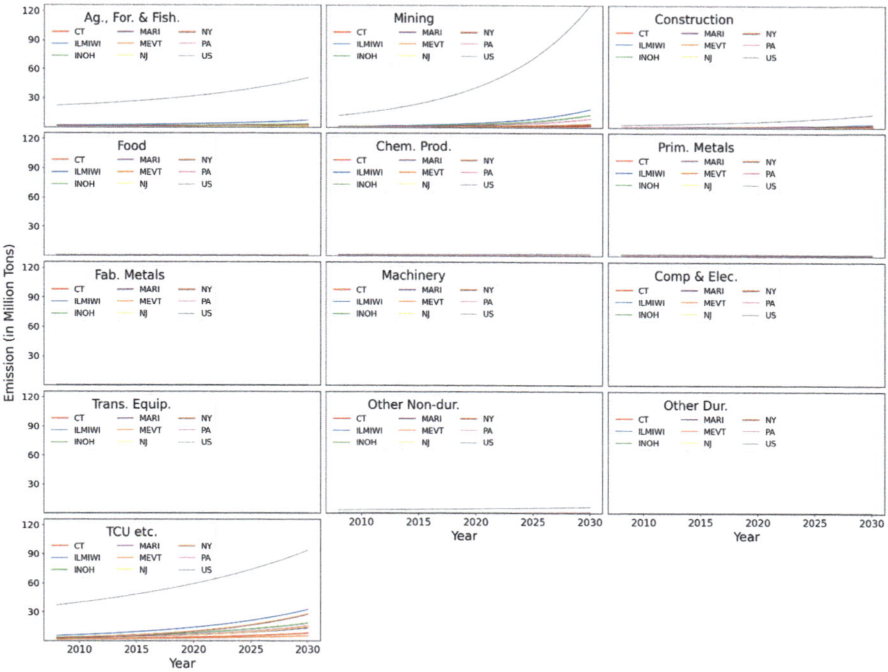

Fig. 8.7 Total environmental footprint of carbon monoxide emissions from production by industry and location

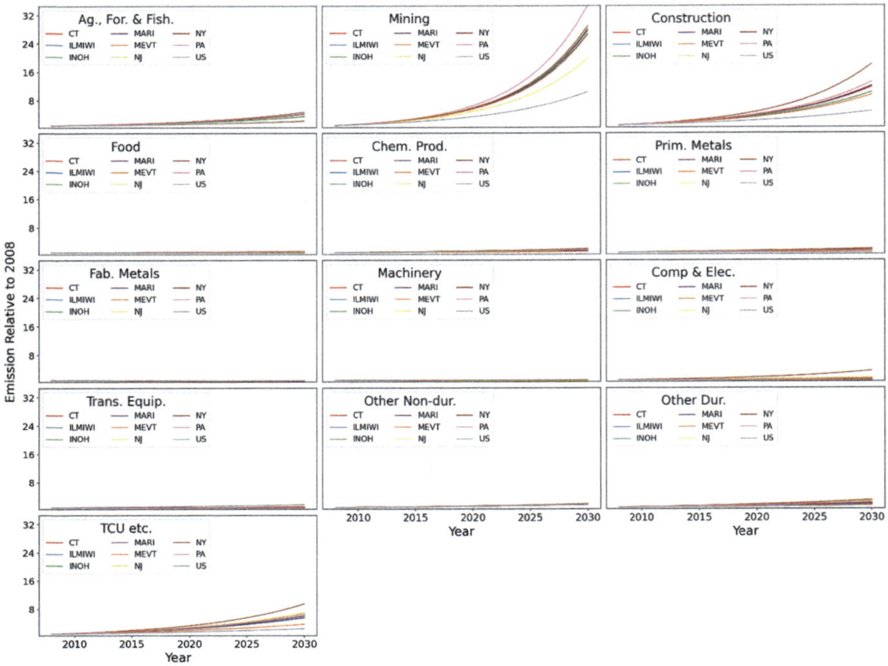

Fig. 8.8 Total environmental footprint of carbon monoxide emissions from production by industry and location relative to 2008

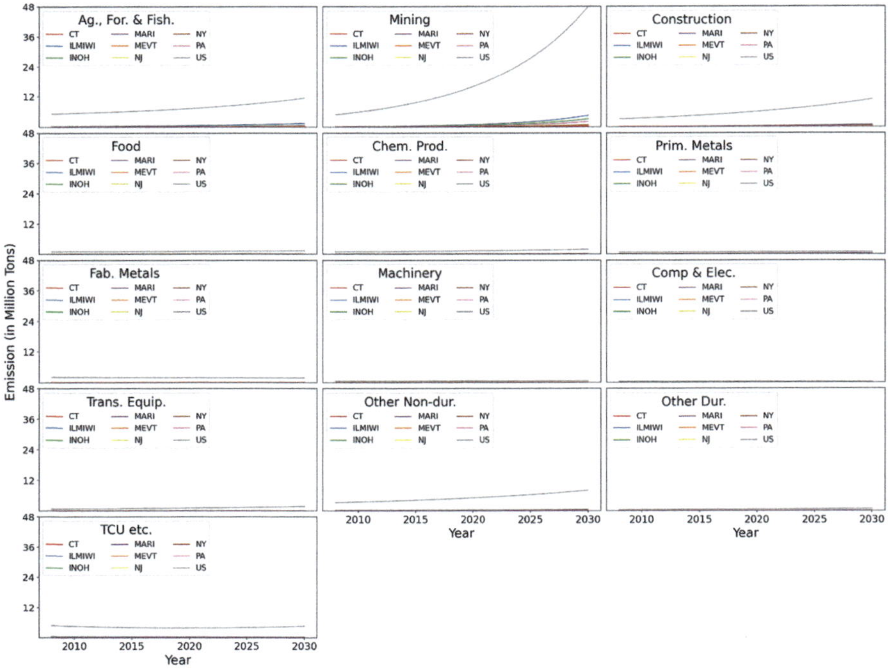

Fig. 8.9 Total environmental footprint of nitrous oxide emissions from production by industry and location

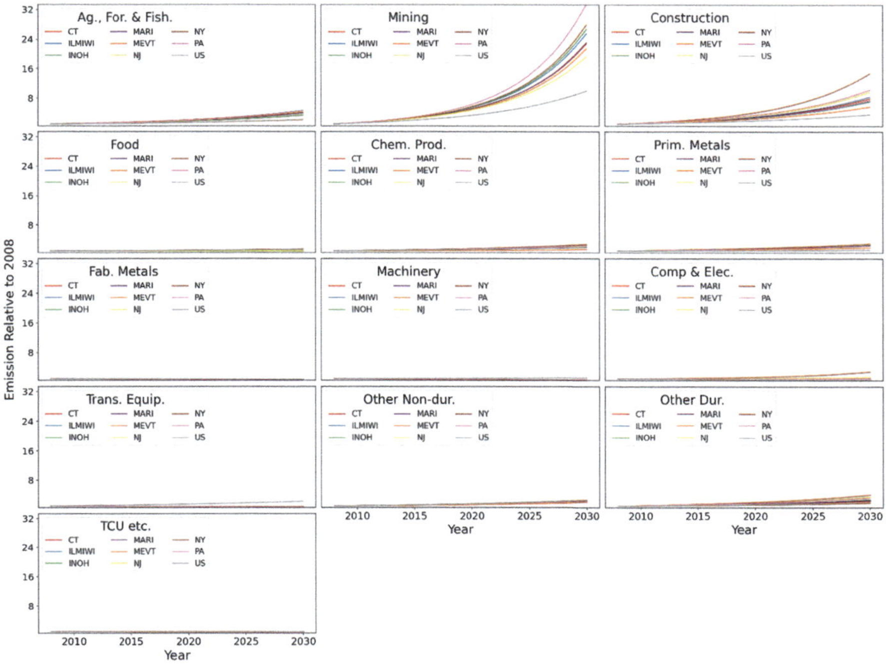

Fig. 8.10 Total environmental footprint of nitrous oxide emissions from production by industry and location relative to 2008

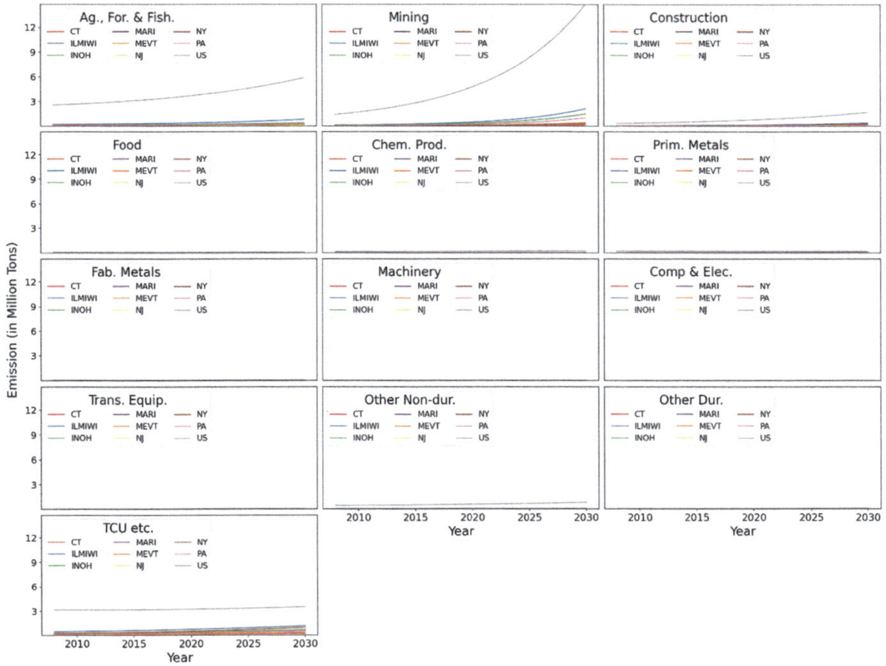

Fig. 8.11 Total environmental footprint of sulfur dioxide emissions from production by industry and location

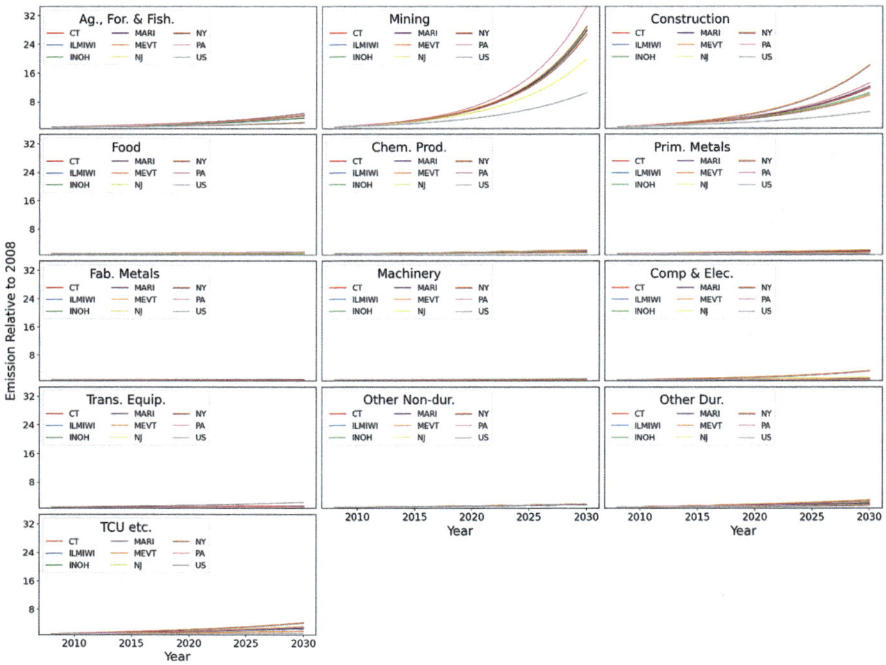

Fig. 8.12 Total environmental footprint of sulfur dioxide emissions from production by industry and location relative to 2008

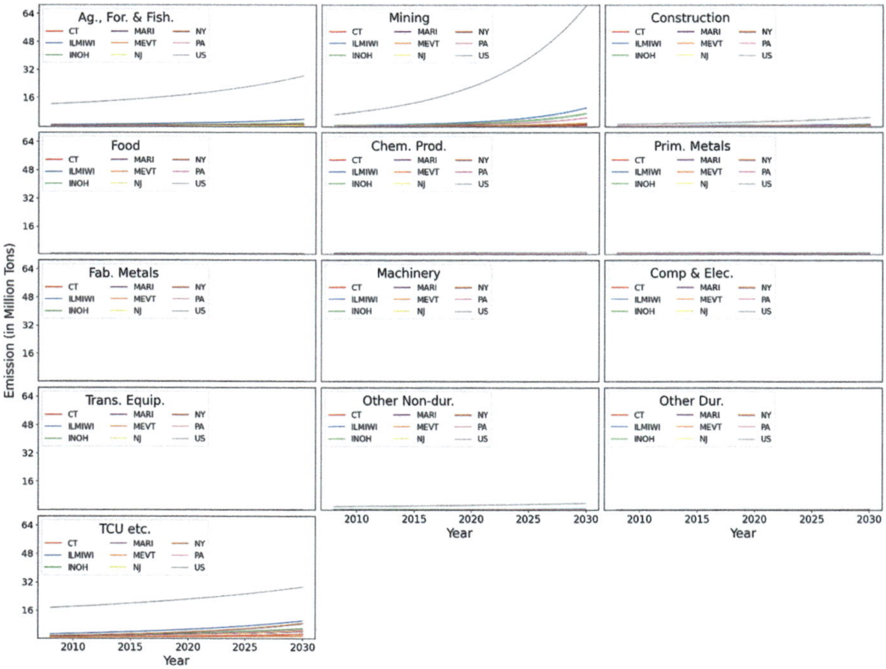

Fig. 8.13 Total environmental footprint of volatile organic compound emissions from production by industry and location

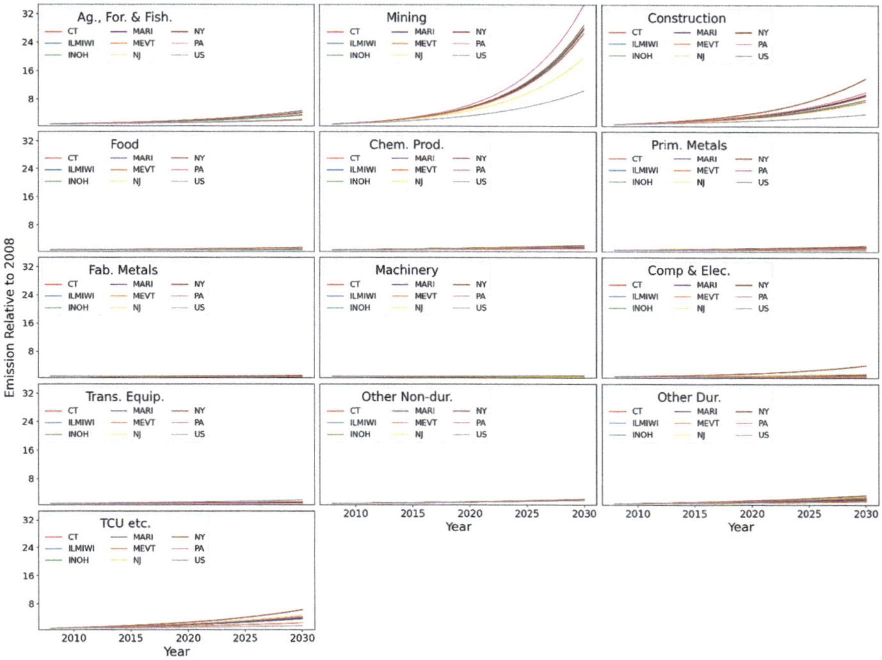

Fig. 8.14 Total environmental footprint of volatile organic compound emissions from production by location and industry relative to 2008

Appendix 8.2 Total Environmental Footprints of Consumption

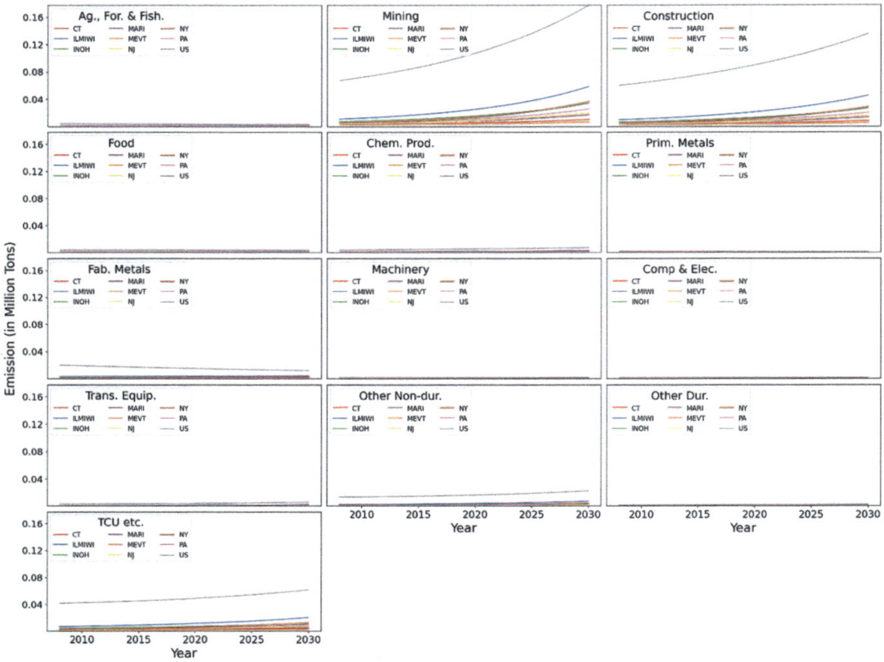

Fig. 8.15 Total environmental footprint of black carbon emissions from consumption by industry and location

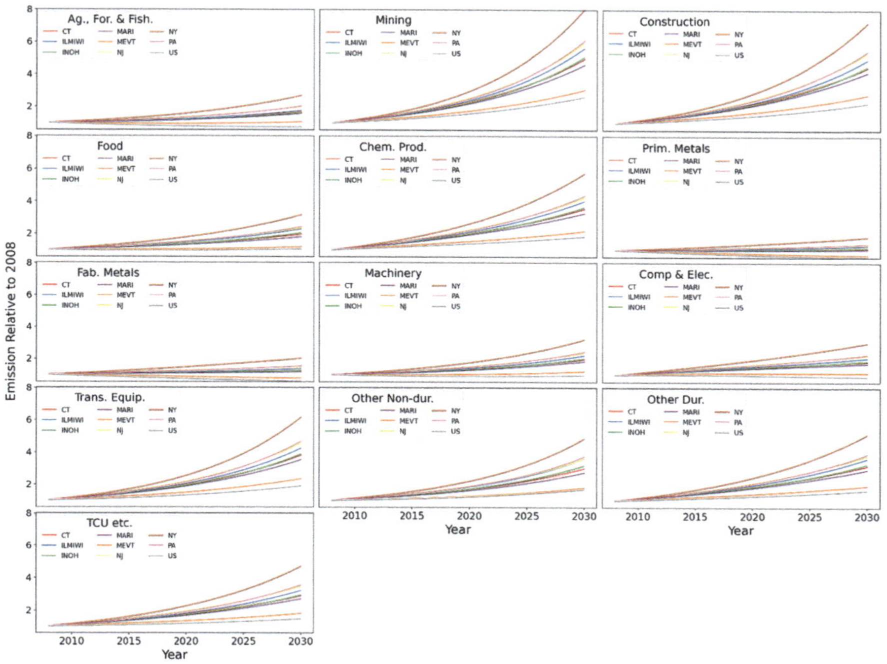

Fig. 8.16 Total environmental footprint of black carbon emissions from consumption by industry sector and location relative to 2008

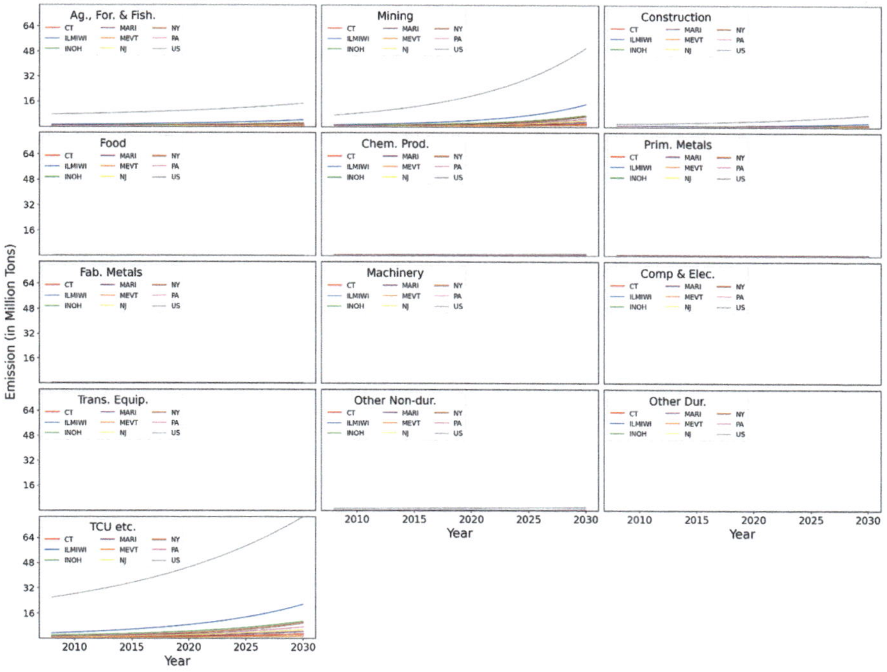

Fig. 8.17 Total environmental footprint of carbon monoxide emissions from consumption by industry and location

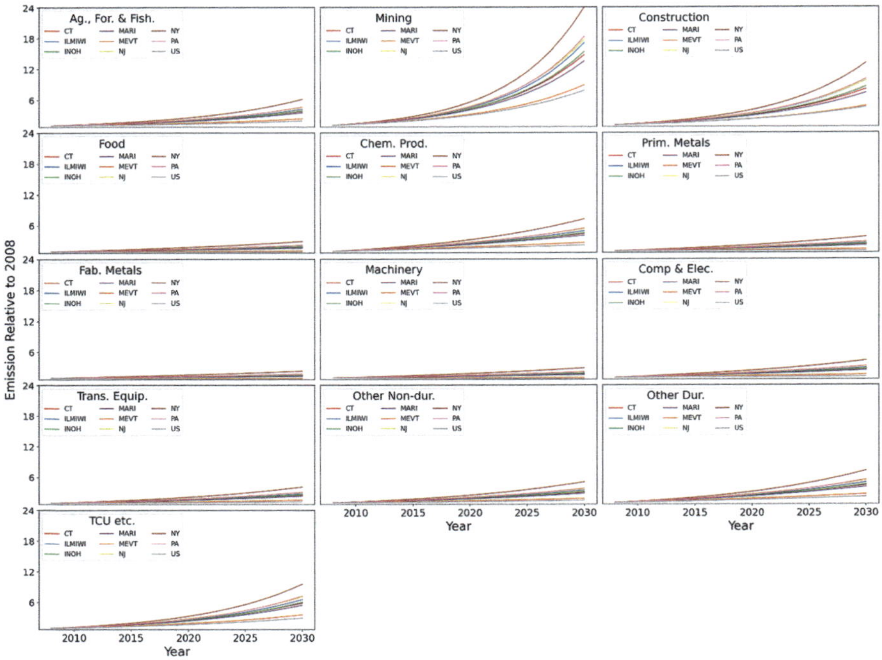

Fig. 8.18 Total environmental footprint of carbon monoxide emissions from consumption by industry and location relative to 2008

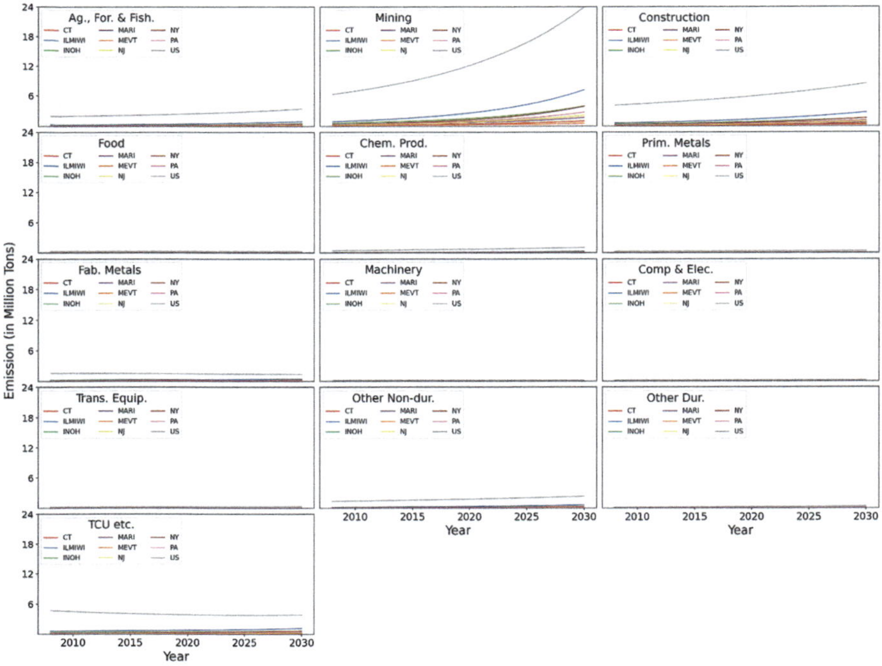

Fig. 8.19 Total environmental footprint of nitrogen oxide emissions from consumption by industry and location

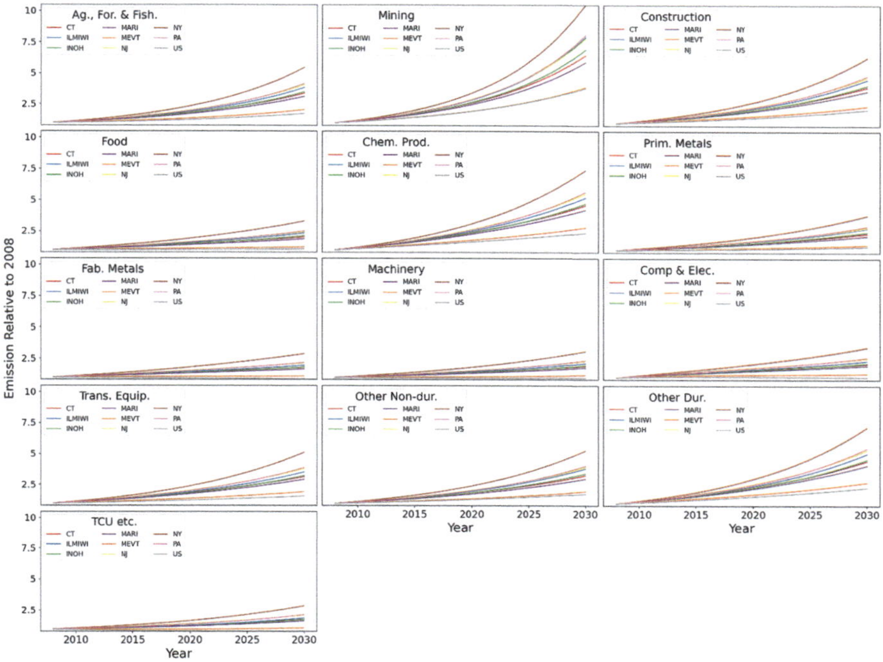

Fig. 8.20 Total environmental footprint of nitrogen oxide emissions from consumption by industry and location relative to 2008

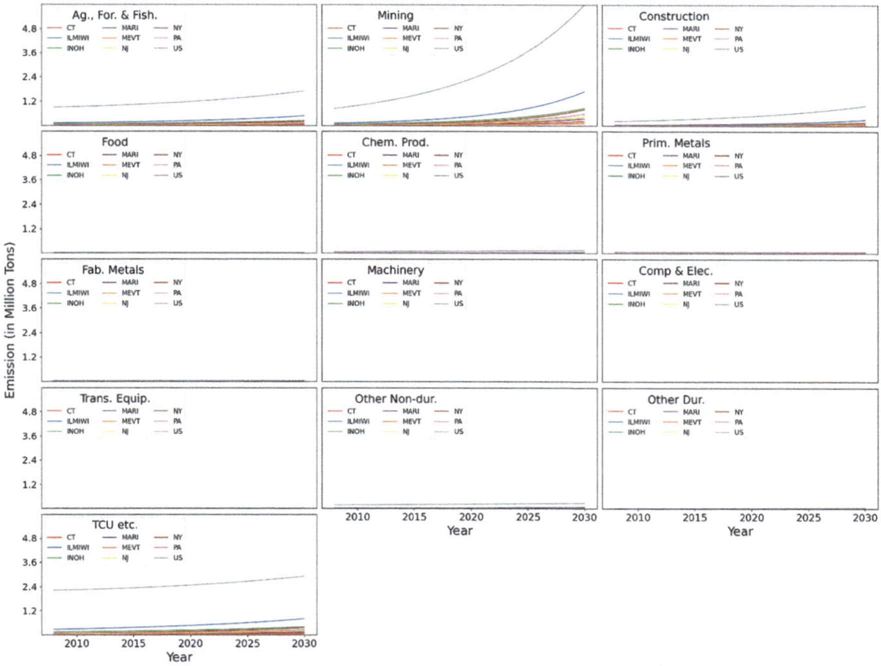

Fig. 8.21 Total environmental footprint of sulfur dioxide emissions from consumption by industry and location

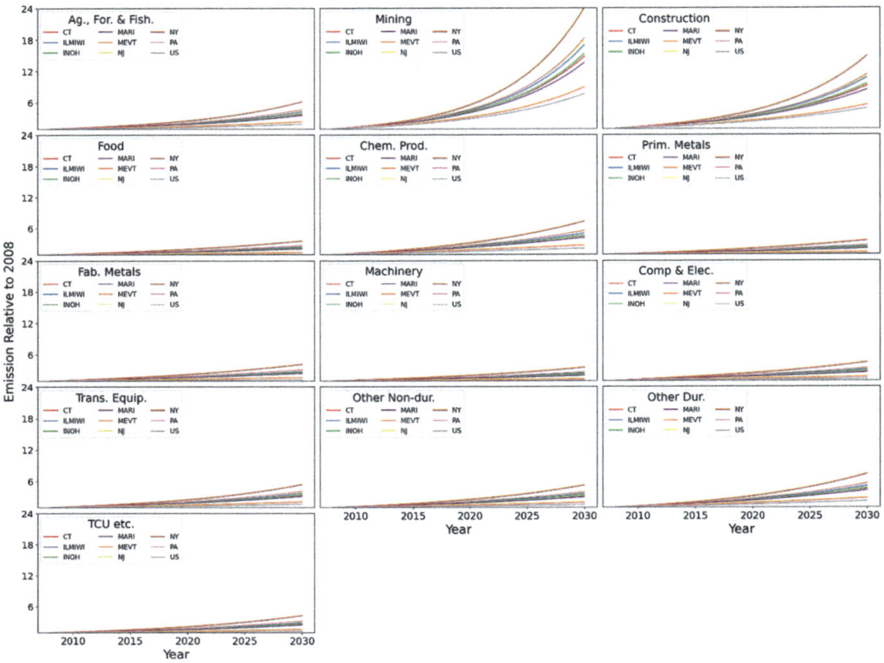

Fig. 8.22 Total environmental footprint of sulfur dioxide emissions from consumption by industry and location relative to 2008

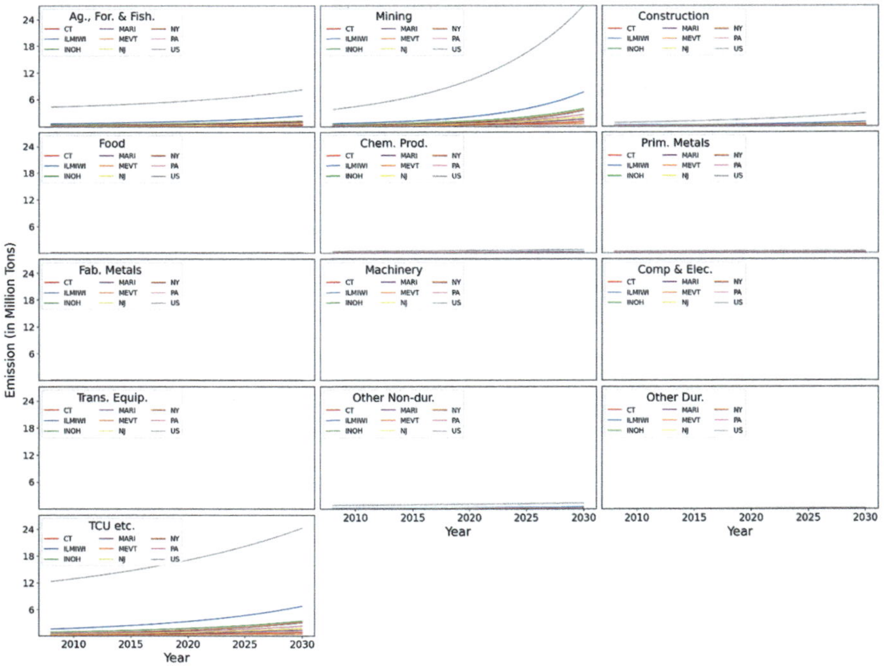

Fig. 8.23 Total environmental footprint of volatile organic compound emissions from consumption by industry and location

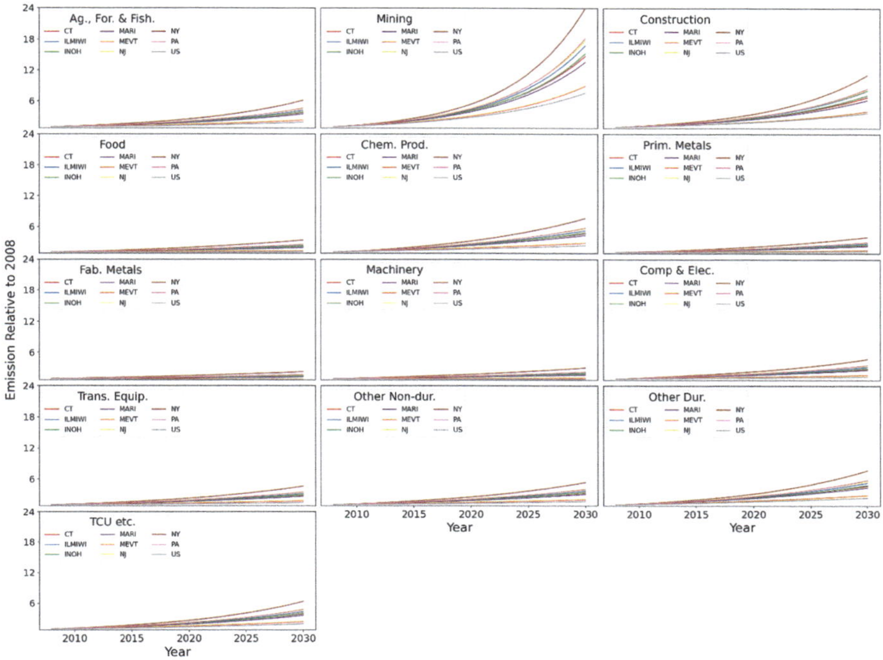

Fig. 8.24 Total environmental footprint of volatile organic compound emissions from consumption by industry and location relative to 2008

References

Brown-Steiner B, Hess P, Chen J, Donaghy KP (2016) Black carbon emissions from trucks and trains in the Midwestern and Northeastern United States from 1977 to 2007. Atmos Environ 129:155–166

Donaghy K, Wymer CR, Hewings GJD, Ha SJ (2017) Structural change in the Chicago region and the impact on emission inventories in a continuous-time modeling approach. J Econ Struct 6:20. https://doi.org/10.1186/s40008-017-0083-x

Ham H, Kim TJ, Boyce D (2005) Implementation and estimation of a combined model of interregional, multimodal commodity shipments and transportation network flows. Transp Res B 39:65–79

Jackson R, Court C (2020) I-O Snap. University of West Virginia, Morgantown

Tao Z, Williams A, Donaghy K, Hewings G (2007) A socio-economic method for estimating future air pollution emissions—a Chicago case study. Atmos Environ 41:5398–5409

Tao Z, Hewings G, Donaghy K (2010) An economic analysis of Midwestern US criteria pollutant emissions trends from 1970 to 2000. Ecol Econ 69:1666–1674

U.S. DOT FHA (2020) National freight transportation trends and emissions. U.S. DOT, Washington, DC

U.S. EPA (1999) National emissions inventory 1999. U.S. EPA, Washington, DC

U.S. EPA (2008) National emissions inventory 2008. U.S. EPA, Washington, DC

Chapter 9
Conclusions and New Directions

Abstract This chapter summarizes the book-length argument developed in the preceding chapters and suggests new directions in which research addressing the coevolution of commodity flows, economic geography, and atmospheric emissions might be pursued.

9.1 Argumentative Summary

In Chap. 2, we reviewed some of the most salient features of globalization and its environmental impacts. We noted in particular that while cities, in which most industrial production occurs, continue to cast a large environmental footprint on their regionally proximate physical environments, they now (after globalization) also exert a strong influence on the natural systems of more remote locations because of their growing interconnectedness and interdependence with other cities and those cities' respective hinterlands. We also surveyed possible responses to these developments and focused in particular on the critical role that urban infrastructure systems can play in mitigating against deleterious environmental effects. And we discussed how one might provide analytical support for managing changes in urban infrastructure systems to achieve such mitigation.

With Chap. 3, our analysis moved to a higher level of geospatial abstraction than individual cities and a narrower interregional focus as we concerned ourselves more with commerce between industrial aggregates in a limited number of states and groupings of states in the Midwestern and Northeastern United States. To examine empirically the coevolution of the three aspects of globalization with which this book has been concerned (commodity flows, economic geography, and atmospheric emissions), spatial time series on interregional interindustry (and intra-industry) sales were needed. Chapter 3 presented and demonstrated a methodology for generating such data. The empirical commodity-flow data reflect 30-year trends of globalization discussed above—the increasing transport intensity of all economic activity, increasing intra-industry trade and the hollowing out of local economies,

K. Donaghy et al., *The Co-evolution of Commodity Flows, Economic Geography, and Emissions*, Advances in Spatial Science,
https://doi.org/10.1007/978-3-030-78555-0_9

and fluctuating nonpoint-source emissions associated with that trade.[1] From plots of intra-industry trade data displayed in Chap. 3, we also observed that changes in these trade volumes were sufficiently large to warrant efforts to estimate a structural-equation model that may suggest explanations of these changes, even as changes over time in interindustry sales coefficients, from which the data were derived, were not dramatic.

In Chap. 4, we analyzed and graphically portrayed the evolution of commodity flows in value terms (constant 2001 dollars) converted to physical measures of freight movement (in ton-miles) and commented on the changes in aggregate volumes of shipments, intra-industry shipments, and patterns of associated black carbon emissions. We found that for some sectors and between some states there is considerable variation in intra-industry trade (IIT). IIT is very limited across all states in some industries in which there is specialization—e.g., transportation equipment—but also fairly high across many states in other industries—e.g., other nondurable manufacturing. Changes in IIT patterns have not been dramatic in the sample period we examined, leading to speculation that much of the growth in IIT may have occurred earlier than 1977.[2] Some regional IIT patterns in the Midwest-Northeast corridor have emerged more narrowly within the Midwest—in the cases of agriculture, food product manufacturing, and TCU, Services, and Government Enterprises—while corridor-wide patterns have emerged in others—in the case of other nondurable manufacturing—and virtually not at all in others—e.g., in the case of transportation equipment.

In Chap. 5, we analyzed in finer detail the relationship between the spatial distribution of black carbon (BC) emissions and their circulation through the Midwestern and Northeastern regions of the United States and the offsetting developments of exponentially increasing freight movement and emissions-reducing technologies. We established a framework for estimating BC emissions from trucks and trains using the data we derived on interregional interindustry sales and converting them to freight shipments. Employing this framework we addressed two questions: (1) what were the changes in BC emissions between 1977 and 2007 and what were the determinative factors driving these trends, and (2) what economic sectors accounted for the largest share of those emissions? In answer to the first question, we found that BC emissions from freight transportation exhibited little trend in the 30-year period, with emissions in 2007 being only 10% less than those in 1977. However, we observed a peak in the early 1980s, when emissions were approximately 40% higher than their 2007 values and another peak in the 1990s, when emissions were approximately 25% higher than the 2007 value. From 1997 to 2007 emissions declined gradually and steadily. We noted that this pattern resulted from the two offsetting developments noted above. Regarding the second question, we found that the fabricated metal sector accounted for more than half of

[1]This volume's penultimate chapter portrays future environmental footprints of production and consumption if these trends should persist.

[2]But see Hewings and Parr (2009) who find more robust evidence of increasing IIT in the Midwest.

nonpoint-source BC emissions in 1977, with its share peaking in 1983 and again in 1996, and gradually decreasing from 1997 to a third of total BC emissions by 2007. Other sectors exhibited dramatically different trends; notably, mining was twin-peaked. In Chap. 5, we observed that changes in nonpoint-source BC emissions were not uniform across the Midwestern and Northeastern United States and that such emissions increased most around urban centers as transportation hubs. Our findings compare favorably with other BC inventories (e.g., MACCity) for urban centers and surrounding (metropolitan) areas. However, our findings show greater BC emissions along transportation corridors than the MACCity inventory, which allocated emissions across space using population shares. We ascribe the differences in subregional trends in BC emissions to regional changes in the production of finished and semifinished goods and the freight distribution system, which connects them, all under the influence of broader scale globalization.

Up through Chap. 5, we have presented various means of tracking the coevolution of commodity flows, economic geography, and atmospheric emissions but still not offered an explanatory model. In Chap. 6, we discussed structural-equation models of commodity flows and demonstrated how features of such models in two distinct lineages—those of spatial interaction parentage and those of interregional input-output parentage—can be integrated in two new specifications that embody recent theoretical developments and extensions that permit examination of different classes of contemporary problems that lie beyond the explanatory reach of previous models.

In Chap. 7, we presented the operationalization and econometric estimation of a modified version of the dynamic continuous-time structural-equation model of commodity flows elaborated in the previous chapter. Much of Chap. 7 was concerned with the development of other regional economic data needed to supplement the commodity flow data already derived.[3] From our estimation work, we found that a model whose specification is based on a New Economic Geography characterization of producer behavior can fit the data very well and capture important stylized facts of the evolution of globalization in the sample period, 1977–2007. While estimates of many key structural parameters were statistically discernible and plausible, and thereby provide support for explanations of the evolution of commodity flows that place emphasis on the role of increasing returns to scale in key industries, hence imperfect competition, cost savings through fragmentation of production processes and increasing intra-industry trade, and declining relative costs of transport leading to increased transport intensity of production and consumption, the large number of parameter estimates that were not statistically discernible suggests that the model specification we have employed can be improved upon. We hope that our making available the data we have used to conduct this analysis will encourage others to take up the challenge to rigorously model the evolution of commodity flows as well as the other side of these phenomena, the

[3]These data are also available to other scholars wishing to pursue the line of inquiry presented in this volume.

evolution of economic geography, and associated evolution of atmospheric emissions.

Chapter 8 recurs to a theme introduced in Chap. 2—that because of the increasing interconnectedness and interdependence of all locations, economic activities of production and consumption in nodal areas of transportation networks have increasingly extensive environmental footprints. In this chapter, we examined this theme and made projections of point-source and nonpoint-source emissions of various pollutants based on a characterization of a spatial economy in which the flows of intermediate inputs at every stage of industrial production and final demand are taken into account. We found that, from the perspective of production, environmental effects (in the form of atmospheric emissions) of increases in production are being offset by improvements in technology leading to cleaner production—even with increasing fragmentation and intra-industry trade in intermediate inputs. From the perspective of consumption, however, the effect of increased production would seem to dominate, if not overwhelm, the technological effect. We also found that changes in emissions due to the increasing demand for energy products, construction, and services (including transportation services) are noticeable from both perspectives of production and consumption.

In concluding Chap. 8, we noted that the extrapolative and conjectural research reported in this chapter can be extended in multiple ways—by widening the geographical scope of the analysis, extending the time series on commodity flows, employing dynamic programming models to forecast the evolution of freight movement, conducting dynamic game analyses with shippers and carriers to assess the impacts of potential regulation of logistics operations, and integrating analyses of the roles of large firms with extensive production networks with sectoral analyses to explore how environmental footprints can be better managed. We hope that other scholars will take up these challenges.

9.2 New Directions

Looking beyond the extensions suggested above, research building on the work presented in this volume might depart in other directions, two of which concern developments in international trade and transitioning to more circular economies.

As this book goes to press, the importance of international supply chains to domestic production has been made clear as automobile and appliance manufacturers in the United States and Europe are temporarily ceasing operations because of their inability to procure sufficient volumes of semiconductor microchips, which are produced in Asia and upon which the operability of vehicles and appliances depend. The criticality of well-functioning international supply chains is also apparent in the dependence of countries around the globe on supply chains to deliver vaccines to protect their populations against an ongoing virulent pandemic. Although the physiology of globalization is changing, as new challenges emerge, production networks evolve and some countries pursue more protectionist policies, globalization itself is still a robustly ensconced phenomenon (Donaghy 2012; The Economist 2021). To understand how domestic commodity flows, economic geography, and atmospheric emissions might co-evolve going forward, the dependence of domestic production networks on international supply chains must be formally represented in models along the lines we have considered in this volume so that the logical implications of changes in supply chain behavior can be examined. There are now data sets that support analyses of such changes. (See, e.g., Baldwin and Lopez-Gonzalez (2015).)

Another important development related to international supply chains with implications for the coevolution of domestic commodity flows, economic geography, and atmospheric emissions is the increasing standardization of transport infrastructures and port facilities. Much of this standardization is being influenced by China's "Belt and Road Initiative" (Wiig and Silver 2019). Wiig and Silver (2019) introduce an innovative conceptual scheme for making sense of the stages through which global infrastructure comes to be configured across urban spaces (the nodes of key transport networks).[4] It may be important to introduce characteristics of transport infrastructure and port facilities explicitly in commodity-flow models to account, going forward, for what gets shipped where and by how.

While reducing atmospheric emissions that contribute to anthropogenic climate change is presently a cause of great concern to most of the world's nations, the need to transition to more "circular economies" is also gaining appreciation.[5] To address the question of *how to promote greater circularity* in production, consumption, recycling, waste handling, etc., spatial-temporal models of "general transport networks" (see the discussion in Chap. 2) will be needed that characterize path dependencies of socioeconomic behavior on infrastructure systems and possible adaptation lags. Given long-lived infrastructure networks whose service lives will surely exceed those of production network configurations (and settlement patterns), such models will be required to help determine *sequences* of spatial configurations of interdependent systems of systems that can lead societies to greater circularity. To be useful in supporting planning initiatives affecting the coevolution of commodity flows, economic geography, and environmental impacts in transitioning to circular economies, modeling of production network behavior will need to be more realistic. We believe that the current state of data availability, high-performance computing,

[4]The steps comprise *speculation* about how infrastructure supports cities' ambitions to become or remain central to international trade, *delineation* of ways in which spaces of standardized infrastructures are configured for such trade, *alignment* through which individual technological, regulatory, or financial components of an infrastructure system are adjusted as necessary to shift the system to new flows and circulations in the global economy, and *pivoting* in which the underlying relationships between cities and the global economy are transformed by implementing a necessary combination of infrastructure projects. (See Wiig and Silver (2019).)

[5]The idea of a *circular economy* is a characterization of how goods and services can be produced and consumed in an ecologically sound and environmentally sustainable manner that meets concerns of overuse of resources, waste generation, and climate change, *inter alia*, through the conscious interlinking of disparate economic activities. (See Donaghy (2021).)

and planning support system development should enable such research to be carried out now.

References

Baldwin R, Lopez-Gonzalez J (2015) Supply-chain trade: a portrait of global patterns and several testable hypotheses. World Econ 38:1682–1721

Donaghy KP (2012) The co-evolution of logistics, globalization, and spatial price competition: implications for a unified theory of trade and location. In: Capello R, Dehntino T (eds) Emerging challenges for regional development: new directions in investments and migration flows. Edward Elgar, Northampton, pp 63–92

Donaghy KP (2021) Getting to a circular growth economy by harnessing circular and cumulative causation. In: Reggiani A, Schintler LA, Czamanski D, Patuelli R (eds) Handbook on entropy, complexity and spatial dynamics. The rebirth of theory? Edward Elgar, Cheltenham

Hewings GJD, Parr JB (2009) The changing structure of trade and interdependence in a mature economy: the US Midwest. In: McCann P (ed) Technological change and mature industrial regions: firms, knowledge, and policy. Elgar, Cheltenham, pp 64–84

The Economist (2021) Message in a bottleneck: global supply chains are still a source of strength, not weakness, April 3, p 9

Wiig A, Silver J (2019) Turbulent presents, precarious futures: urbanization and the deployment of global infrastructure. Reg Stud 53(6):912–923

Index

K. Donaghy et al., *The Co-evolution of Commodity Flows, Economic Geography, and Emissions*, Advances in Spatial Science,
https://doi.org/10.1007/978-3-030-78555-0

Printed by Printforce, the Netherlands